The Helmholtz Curves

forms of living

Stefanos Geroulanos and Todd Meyers, *series editors*

The Helmholtz Curves

Tracing Lost Time

Henning Schmidgen

Translated by Nils F. Schott

FORDHAM UNIVERSITY PRESS

NEW YORK 2014

This work was originally published in German as Henning Schmidgen, *Die Helmholtz-Kurven: Auf der Spur der verlorenen Zeit* © 2009 Merve Verlag Berlin.

The translation of this work was funded by Geisteswissenschaften International—Translation Funding for Humanities and Social Sciences from Germany, a joint initiative of the Fritz Thyssen Foundation, the German Federal Foreign Office, the collecting society VG WORT, and the Börsenverein des Deutschen Buchhandels (German Publishers & Booksellers Association).

Fordham University Press has no responsibility for the persistence or accuracy of URLs for external or third-party Internet websites referred to in this publication and does not guarantee that any content on such websites is, or will remain, accurate or appropriate.

Fordham University Press also publishes its books in a variety of electronic formats. Some content that appears in print may not be available in electronic books.

Library of Congress Control Number: 2014930561

Printed in the United States of America

16 15 14   5 4 3 2 1

First edition

CONTENTS

Lost time is the temporal gap between sensation and movement, perception and thought, decision and action. It is not a time that is forgotten. It is not the time of missing memories. Nor is it a time squandered or wasted. Lost time is never the time of futility, idleness, or sleeplessness. It is the time of suspension, of opening, of possibility.

In the summer of 2009 I was working in the archives of the Paris Academy of Science, doing research for a larger study on the history of brain-time experiments in the nineteenth and twentieth centuries. It was there and then that the Helmholtz curves interrupted my stream of thought. What I suddenly had before my eyes were writing-images that oscillated in fascinating ways between transparency and obscurity. The Helmholtz curves immediately captured me, disrupted the work on my larger study, and drove me to write this book. The rest of my work came to a halt. It had to pause. In this sense, the present book owes its existence to a kind of "lost time."

More than ten years ago, I worked as a clinical psychologist. My main task consisted in placing psychiatric patients in front of computer screens. I instructed them to respond to visual and acoustic signals by quickly pressing buttons or keys. Or I would give them a pencil and confront them with a form on which they had to mark specific letters or characters. Armed with a stopwatch, I tried to determine the time they needed for this task.

A decade later—having in the meantime turned into a historian of science—I found myself in the Paris archives, in a similar situation but with the roles reversed. Now it was I who was sitting at a table with a form in front of me. Magnifying glass in hand, I was waiting with focused attention for what was to come.

Are there reaction time experiments in which the propagation of the stimulus takes more than a hundred years? Are there response times whose length is measured not in tenths and hundredths of a second, but in weeks and months? Are there reactions that require more than 300,000 strokes on a keyboard?

This book, the very fact that it exists, answers these questions. I could not have written it without the generous support I received from many sides. First and foremost, I would like to thank the staff at the archives, museums, and libraries, who granted me access to their collections, especially the archives of the *Académie des sciences* in Paris, the *Universiteitsmuseum Utrecht*, and, in Berlin, the archives of the *Berlin-Brandenburgische Akademie der Wissenschaften* (BBAW), the *Staatsbibliothek*, the *Universitätsbibliothek* of Humboldt University, and the library of the Max Planck Institute for the History of Science (MPIWG).

Thanks to a collaboration between the Virtual Laboratory at the MPIWG and the archives of the BBAW, a representative selection of documents from Helmholtz's papers (letters, manuscripts, notebooks) has been digitized and made available online, at <http://vlp.mpiwg-berlin.mpg.de/library>. In addition, the Virtual Laboratory contains most of the publications by Helmholtz, du Bois-Reymond, Marey, and others quoted in this book. I would like to thank the students, assistants, and colleagues involved in digitizing these sources, especially Kaja Kruse, who also prepared the image files for the present volume.

Since my discovery of the Helmholtz curves, I have had occasion to discuss my project with numerous friends and colleagues. For friendly encouragement, open answers, and subtle reluctance, I would like to thank Timothy Lenoir, Bernhard Siegert, and Hans-Jörg Rheinberger, as well as Bernhard Dotzler, Robert Brain, David Cahan, and Norton Wise. My reconstruction of Helmholtz's time experiments would no doubt have been much more fragmented had I not had the opportunity to discuss my ideas with the open-minded historians of physics at the MPI, and I would like to thank Jochen Büttner, Shaul Katzir, and Christian Joas in particular. For comments, criticisms, and corrections, I thank all those who agreed to read a chapter of the book or even the entire manuscript: Matthias Flaig, Jan Bovelet, Gabriel Finkelstein, Dieter Hoffmann, Manfred Laubichler, Christian Reiss, and Skúli Sigurdsson.

I also thank the University of Regensburg for supporting this project.

Finally, I would like to thank Nils F. Schott for his meticulous work on translating and editing the text, and I would like to express my gratitude to Todd Meyers, Stefanos Geroulanos, and everyone else at Fordham University Press, whose commitment and enthusiasm have brought this book to life.

The Helmholtz Curves

Introduction

> To think time is to place life in a framework.
>
> —GASTON BACHELARD[1]

At the beginning, two images. Both were created in the middle of the nineteenth century, both of them are signed "Helmholtz," and both are movement-images as well as time-images. Our first look at them is deeply anachronistic. They appear to be horizontal filmstrips or elongated negatives of black-and-white photographs. Around 1850, however, such a use of celluloid was only a remote possibility. Not until the 1880s was celluloid turned into the quintessential storage medium for photographic and cinematographic images. And yet, our two images already deal with cinematics.

Both are carefully mounted on white cardboard. Mounted that way, scratches and spots become visible, as if the dark rectangles had been manipulated from behind with a needle. The first image (Figure 1) shows three rolling lines. Arranged one above the other, they are reminiscent of ocean waves gradually flattening on a beach.

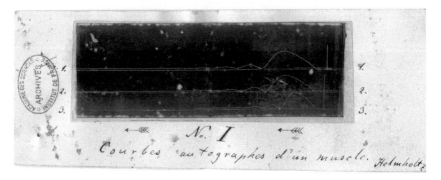

FIGURE 1. Helmholtz Curves No. I, "Autographic Curves of a Muscle" (1851). Reprinted with permission from Académie des sciences—Institut de France, Paris.

The second image (Figure 2) takes us from the beach out onto the high seas. We see what appears to be a swell, spread across three levels. Unlike the rolling lines in the first image, these long arcs are drawn twice. Calmly, they run in parallel, three times, with tiny gaps between them.

Despite the significant differences between the two images, they show the same phenomenon. Every observer has already experienced it: the twitching of a muscle. But hardly anyone has ever seen it in this form, and an observer in the middle of the nineteenth century certainly had not. In fact, it took a complicated and fragile *drawing machine* to produce these recordings of movements, a heterogeneous and precarious assemblage consisting of frog muscles, frog nerves, batteries, a rotating cylinder, a stylus, a layer of soot, and a host of other components.

The lines and arcs registered on the small transparencies are as closely attached to the materiality of this machine as a decal is to its backing paper. They really stick to it. On the one hand, the machine is responsible for the obvious difference between the images: The recording of the same process produces a steep wave in one case and a flat swell in the other because of the different speeds with which the movement of the muscle was registered. On the other hand, these movement-images are time-images for this very reason. If we relate the size of the recorded lines and arcs back to the dimensions and the rotational performance of the machine, we can calculate

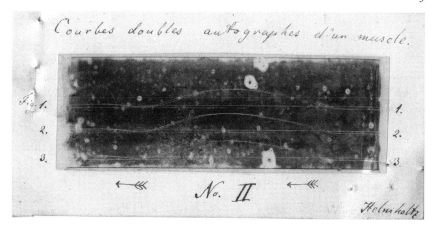

FIGURE 2. Helmholtz Curves No. II, "Autographic Double Curves of a Muscle" (1851). Reprinted with permission from Académie des sciences—Institut de France, Paris.

the exact time during which the drawing occurred—or did not occur. In other words, what is *invisible* is just as important as what appears as *visible*.

In the first image, motion-wave and resting-beach are initially one. At the beginning, they draw a single, shared trace. The muscle in the machine was stimulated. But for a short moment, it did not think (as it were) to contract or to draw.

In the second image, the one with the swell, the crucial element is the dark gap between the parallel arcs. The interstice corresponds to the distance, in two successive experiments, between different points at which the nerve was stimulated to make the muscle twitch. The greater the distance between the nerve point and the muscle, the larger the empty darkness between the arcs. It is this darkness that corresponds to the time that has been "lost" in the nerve.

Surveying and measuring

Our two images date from 1851. They thus provide very early evidence for the process of "picturing time" that historians of science have shown to be characteristic of the experimental life sciences since the last third of the nineteenth century.[2] Put differently, Helmholtz's curves anticipate in striking fashion the famous motion studies conducted by Etienne-Jules Marey in the 1870s and 1880s. And there is more. With a sharpness that has survived to this day, the curves fixed in these images mark the beginning of a new age, an age that sought to quantify the living body.

Around 1850, the activity of surveying the world increasingly moved from the vastness of landscapes and the entire globe to the delimited regions of the laboratory. Simultaneously, the focus of survey activities shifted from the spatial to the temporal. The existing bio-/geography was transformed into a rapidly evolving innovative chronography. More and more, surveys concerned the inside of the body. They literally got under the skin.

By the same token, the very meaning of surveying and measuring (*vermessen* and *messen*) was transformed. Around 1800, at the heyday of Alexander von Humboldt's career, it was sufficient to observe the needle of a surveying instrument to believe that one was thereby looking "into the interior of the world."[3]

A mere two generations later, researchers had done away with this system of representation. Helmholtz and other physiologists had replaced it with a regime of experimentation that appears, by comparison, radically modern. In this regime, *what was measured was produced by the measurement.* Ultimately, this applied even to the separation between the inner and the outer world. In fact, the experimentalists on the ascent were no longer interested in surveying the known, the tangible—or at least attainable—in accurate and ever more exact ways. Instead, their project was to capture the unknown and, in a sense, to produce the hidden interior by means of measurements on, around, and in the bodies of the living.

This was no longer mere "physiology": It was just as much an "organic physics," as the historical actors put it.[4] Perhaps it was even a "physio-technics," as we might call it in allusion to Bachelard's concept of a "phenomeno-technics." At any rate, the relatively concrete act of surveying was transformed, around 1850, into a much more abstract activity of measurement. This activity no longer conjured the horror of things. It did something completely different. It brought new things to light, things that were horrifying as well as non-horrifying, unexpected as well as foreseeable, clear and crisp on the one hand, formless and full of flaws on the other. Since that time, laboratory phenomena have been strictly contemporaneous with their being measured. Their history, accordingly, has to be written as a history of these measurements.[5]

The Helmholtz curves provide ample evidence for this transition to a modernity of measurement. In 1847, wave writers, so-called kymographs, had been introduced into the practice of experimental physiology.[6] The aim of this technical innovation was to record the specific dynamics of vital phenomena such as circulation or respiration so they could be observed and studied in greater detail. Fewer than five years later, Helmholtz was no longer concerned with registering muscle contractions as a further manifestation of life in its own language (in the *"langage des phénomènes eux-mêmes,"* as French physiologists would put it in the 1870s in describing the "graphical method").[7] His experimental practice no longer focused on the form of the wave drawings, which for Marey would still be so "striking."[8] The crucial element now was the obvious possibility of grasping and accurately determining a temporality within the living body that otherwise remained largely hypothetical.

This is why two images are posted at the opening of this study. The first image stands for the traditional scientific demand to communicate the results of accurate surveys (*Vermessungen*). The second stands for the avant-garde requirement to also represent what had been the subject of measurement (*Messung*).

The diagram of an experiment

The Helmholtz curves offer insights into the early practices of the experimental life sciences. We might liken the stage of these practices during Helmholtz's time to the early stages of manufacturing prior to the advent of large-scale industry. In the middle of the nineteenth century, there were no laboratories for physiological research in the modern sense, and experimental setups often had to be assembled by putting together instruments and procedures adapted from other areas. Nor was collecting physiological data by any means a factory-like business; it was a matter of working with individual hands and eyes. Journals for physiology did exist; text genres such as the abstract, however, were still unknown. Beside English, the language of science still branched off into Latin, German, French, and Italian. And instead of a single committee of a foundation based in Sweden dominating all other voices, there were competing national academies of science, eagerly evaluating and honoring scientific excellence.

Despite and apart from this gradually emerging system of scientific research, contemporaries were soon convinced that the research behind the Helmholtz curves—his measurements of the propagation speed of stimulations in nerves—had to be considered a "classical study."[9] In fact, Helmholtz's measurements were rapidly accepted as exemplary achievements of the modern life sciences, and the methodological standards they had set were recognized to be authoritative. Helmholtz's time experiments accordingly found their place as landmarks in standard accounts of the history of the emerging disciplines of modern physiology and psychology. In 1969, the physiologist Charles Marx even went so far as to present Helmholtz's time measurements as the first new data in neurobiology since antiquity.[10]

Recent studies in the history of science consider Helmholtz's measurements to be among the exemplary experiments of modern (life) science. Frederic Holmes and Kathryn Olesko, for example, have argued that the historical significance of Helmholtz's psychophysiological time experiments and the curves related to them can be compared to the importance of Lavoisier's first studies on the composition of air, Mendel's famous hybridization experiments, and the fabulous Meselson-Stahl experiment on the semiconservative replication of DNA. According to Holmes and Olesko, Helmholtz's research provided a general "image of precision" that was of key importance for the experimental life sciences of the late nineteenth and early twentieth centuries.[11]

At the same time, Helmholtz established a specific kind of experimental practice that was continued well into the twentieth century and broke new ground in neurobiology, experimental psychology, and brain research. From Franciscus Donders and Wilhelm Wundt to Keith Lucas, Hans Berger, and Benjamin Libet, outstanding scholars used experimental time measurements and curve drawings to bring light to the dark chambers and tubes from which our thinking, feeling, and doing emerge every day.[12]

These experimental time measurements also had a considerable impact on philosophy. In his two volumes on the cinema, Gilles Deleuze describes the years around 1900 as a "historical crisis of psychology."[13] As Deleuze explains, this crisis, crucially, was based on and referred to a precarious relationship between image and movement. Because of "social and scientific factors,"[14] which culminated in the emergence of the cinema, the traditional duality—image without movement on the one hand, movement without image on the other—had become untenable. Around the turn of the century, more and more movements invaded conscious experience while more and more images entered the material world. According to Deleuze, this resulted in new kinds of problems: "How is it possible to explain that movements, all of a sudden, produce an image—as in perception—or that the image produces a movement—as in voluntary action?" According to Deleuze, Bergson deserves credit for responding to these problems with a new conception of the brain. Bergson superimposed image and movement onto one another such that the brain became a *gap*, a hesitation and delay within a world of universal mutability: "The brain was now only an interval [*écart*], a void, nothing but a void, between a stimulation and a response."[15]

In what follows, I will show that since the middle of the nineteenth century, physiologists such as Helmholtz established and promoted this remarkably empty conception of the brain. They did so, however, not primarily by way of theories or a philosophy but by means of an experimental practice. In fact, the Helmholtz curves mark the beginning of an extended lineage of psychophysiological research machines that, in different contexts and by making use of different instruments and procedures, explored the gap between stimulation and contraction, the interval between stimulus and response, and the discontinuity between sensation and movement.

The accurate and reliable delimitation of these interstices was a decisive prerequisite for conducting studies to analyze the functioning of the brain and the nervous system long before CT and PET scans promised to show the mind at work. All that seemed necessary for physiological analyses were experimental variations and subtractions that drew comparisons between the "intakes" and "outputs" of an organism—similar to the research practices of organic chemistry that were already well established in the 1840s and 1850s. In other words, precision time measurements allowed for obtaining scientific knowledge about the functioning of the brain and the nervous system even "without an acquaintance with the apparatus" (as chemist Justus von Liebig had put it),[16] that is, without any concrete knowledge of neurons and synapses. The productivity of analytical time experiments on the boundaries between physiology and psychology or between the natural and the human sciences was due to exactly this kind of black-boxing.

However, this is only part of the argument that the present book develops, drawing on both Deleuze *and* Bergson. When Bergson describes the brain as a kind of telephone exchange that "allows communication or delays it,"[17] he alludes, intentionally or not, to the fact that the scientific practices in question would not have been possible without the industrialization of communication that had started in the 1830s. Before there was the telephone, there was the electric telegraph to provide the infrastructures decisive for "delaying." This concrete connection between physiology and emerging networks of communication and experimental setups— the interrelation between brain, time, and machine—is at the center of this book.

The Helmholtz curves, however, do not just mark the beginning of a new lineage of research machines. They also stand for an end, the attainment of a goal. At least from Helmholtz's perspective, these curves mark the (preliminarily) last stage of a process he had started in Potsdam in 1848 with the construction of a "frog drawing machine."[18] The following year he took up, continued, and further developed this research process by means of galvanometers and telescopes in Königsberg, the very place where, in 1851, he would bend it back, as it were, onto machinic curve drawings. The very form of this process, which is itself a kind of curve, shows clearly that Helmholtz's time experiments neither gave a definitive answer to a well-articulated question nor led him, in systematic and planned ways, to a discovery.

Quite the contrary, even. On the one hand, Helmholtz's time experiments often present themselves as a surprising succession of forward, backward, and sideways movements that almost constantly alter his research questions and the topics he is interested in. On the other hand, at the open end of these progressions and transgressions, we do not simply encounter a new scientific fact. We are presented with a scientific *and* technological feasibility. Thanks to their changing materiality and semioticity, Helmholtz's research machines in fact devised an experimental *diagram* that indicated a solution of scientific problems that were yet—and are still—to be concretely articulated.

Drawing a diagram is, however, different from delimiting a territory. In fact, Helmholtz's time experiments did not define any circumscribed or fixed area of research. Instead, they highlighted oppositions or stationary points between which subsequent machines for psychophysiological time research would oscillate. First, in terms of scientific disciplines: Helmholtz's investigations move between biology and physics, between animal physiology and human psychology, between the study of muscle movements and the stimulation of nerves—that is, motor nerves on the one hand, and sensory nerves[19] on the other—and they finally turn to cerebral processes. Accordingly, there are significant changes in the biological components employed in his experimental setups: on the one hand, prepared muscle samples and nerve-muscle samples taken from frogs, on the other hand whole human beings, male as well as female. Second, in semiotic terms, too, Helmholtz's experiments display polarities: The curves are only *one* product of his machine work. The other product consists in numerical values obtained by measurement, that is, in numbers, numbers, numbers. While the indexical signs (the curves) mostly served to explore and visualize, the symbolic signs (the numbers) referred to the ideal of precision, to the accurate measurement of time. Image and logic, icon and number, *Bild* and *Zahl*, therefore, do not yet belong to separate traditions, as in the laboratory practice of physicists involved in big science.[20] Instead, they figure as closely related aspects of a single experimental process in the emerging field of physiology. Third and finally, Helmholtz's experiments also move between cultural polarities: They were carried out on the periphery of the Prussian research landscape, in Potsdam and Königsberg. Yet the knowledge that these experiments incorporated and produced connected them with the metropolitan spheres of scientific and technological activity, with Berlin as well as Paris.

Modernity, Communication, and Control

This book, opening with images of the Helmholtz curves, aims to reconstruct the inner tension and dynamics of the experimental practice that produced them. The argument developed throughout is that the constitution of this practice cannot be separated from the contradictory temporalization processes that characterized the nineteenth century. Often described as the age of acceleration and speed, that century saw Romanticism's relaxed attitude toward time and the French Revolution's new political calendar replaced by a more objective, linear, and dynamic sense of time. Wolfgang Schivelbusch, Stephen Kern, Paul Virilio, Peter Weibel, and others have powerfully shown how the advent of steamships, railroads, telegraph lines, newspapers, and journals, as well as the increasing use of machines and power plants in industrial production, resulted in profound changes in the human experience of time. These changes were experienced, on the one hand, as an intensification and acceleration and, on the other, as a shrinking of space and an alienation of human nature.[21]

Helmholtz's time experiments were closely connected with these temporalization processes, especially in technical terms. The research machines Helmholtz developed in the period between 1848 and 1851 to investigate the activity of muscles and nerves relied on two central acceleration technologies of the nineteenth century, the steam engine and the electrical telegraph. Contemporaries in the 1850s were already aware of the connections between Helmholtz's experimental technique and the steam engine. At least to his fellow physiologists, it was obvious that the curve method was modeled on the production of so-called indicator diagrams, a procedure developed by James Watt and John Southern to register the work performed inside steam engine cylinders. Recent studies in the history of science have underscored the importance of this kinship. Robert Brain and Norton Wise, in particular, have shown that Helmholtz's psychophysiological time research was carried out in a technological and economic context marked by the cult of the (steam-powered) machine and the lure of antiquity, a context in which skill in drawing, bodily exercises, and the will to construct were meant to form an urban, harmonious whole.[22]

Building on these findings, I will argue that the Helmholtz curves do indeed refer to an environment of technology and economics. I will show, however, that they also point to an internal milieu of experimentation. The curves can hardly be understood if they are studied in isolation from the precision measurements Helmholtz performed earlier using a different technique, the so-called Pouillet method. This method, however, tied to electromagnetism and to telegraphy in particular, has nothing to do with steam and other engines. Writing about Helmholtz's later studies in physiological optics and acoustics, Timothy Lenoir has emphasized the importance of media technologies for physiologists' experimental practice.[23] This argument will be reiterated here with respect to Helmholtz's time experiments.

In particular, I will show that it was not only physicists such as Claude Pouillet who, around 1840, began to apply electromagnetism to the problem of precision time measurements and the quasi-instantaneous communication of time-related data;[24] they were joined by pioneers of telegraph technology, among them Louis Breguet (a grandson of Abraham Louis Breguet, the famous watchmaker from Neuchâtel), Charles Wheatstone, and Werner Siemens. In this context, two areas of potential usage repeatedly came up. The first was measuring the velocity of projectiles for military purposes, the second centralizing control of networks of electrical clocks.

In his time measurements, Helmholtz both relied on these technologies and projects and opened up a new field of application for them, namely physiological research. On the level of discourse, Helmholtz's points of reference were the lectures and articles published by Siemens and other telegraphy pioneers. On the level of material and semiotic practices, Helmholtz connected, above all, with the electric telegraph developed by Carl Friedrich Gauß and Wilhelm Weber in Göttingen in the early 1830s. That early telegraph relied on the principle of the torsion balance, the same principle on which Gauß and Weber also based their measurements of terrestrial magnetism. In daily routines, the telegraph served, among other things, to synchronize the clocks in the two observation stations of the Göttingen Association for Terrestrial Magnetism. In other words, the telegraph was the technological nucleus and germ cell for the international network of scientists Gauß and Weber started to establish in 1836. At the same time, their telegraph provided the model for the chronometer Helmholtz would use in his experiments. As we will see, Helmholtz's physiological precision timer could also be thought of as a torsion balance and, as a consequence, could be handled with similar precision.

Yet Helmholtz relied on telegraphy not just at the laboratory bench. He also alluded to the technology in his writings. By making reference to the electric telegraph, Helmholtz illustrated the interplay of brain, nerves, and muscles as a sending and receiving of messages within the body. At this point it becomes quite explicit that the experimental process in question is not only connected with the modern world of energy and labor but also, with equal importance, is linked to the modern world of communication and control.

There is another important aspect in which Helmholtz's experiments are tied to the acceleration processes of the nineteenth century: the formats he chose for transcribing and publishing his results. Since the beginning of the eighteenth century, the world of scientific journals had been growing continuously. Yet the first third of the nineteenth century saw a veritable explosion in the number of review journals and annual reports. These facilitated and thereby also accelerated processes of communication among an increasing number of scientists. In the 1840s, the experimental life sciences were drawn into these new dynamics of publishing. Subsequently, the number of physiological "discoveries" that were announced in scientific journals and in the bulletins of learned societies (academies, for example) constantly increased, in contrast to original research published in monographs.[25]

Helmholtz was eager to take advantage of this emerging infrastructure for communicating the results of his own research as efficiently as possible. In January 1850, for example, he published a preliminary report in which he listed the first results of his time measurements. A few months later, he followed up with an extended magazine article containing detailed protocols of his experiments. In addition, the initial report was widely reprinted in academy bulletins and relevant scientific journals. Translations of Helmholtz's articles also appeared in foreign periodicals, especially in France. Given this prolific series of publications in the years from 1850 to 1852 (see the chronological bibliography at the end of this book), the impression arises that the extremely short times Helmholtz had measured also suggested extremely fast ways of communicating these findings. Issues of scientific priority were in fact at stake—especially once Helmholtz went beyond the confines of the German-speaking world.

The quick succession of Helmholtz's publications sheds light on a classic theme in the sociology of science. As is well known, the establishment and acceptance of scientific priority is tied in to the distribution of symbolic capital. However, it also ties in to an economy of time—and history.[26] Given the acceleration process orchestrated by steam engines and telegraphs, it does not appear to be an accident that the increase of publications in scientific journals led to concrete endeavors in writing the history of science. Rooted in the same contexts in which Helmholtz was performing his experiments, these projects quickly went beyond simple eulogies and commemorations. They also went beyond the plain text format. The Berlin physicist Johann Christian Poggendorff, for example, illustrated his data concerning human actors and their discoveries in the history of science with graphical "life lines" (*Lebenslinien*).[27]

Slowing down

As familiar as the combined image of acceleration and modernity might be to our conception of the nineteenth century, the century was also a period of slowness and hesitation. In Walter Benjamin's view, at least, the rapidly growing cities of the time were not just places of hectic productivity. They were also vast landscapes in which *flâneurs* would wander through arcades, taking their turtles for a walk.[28] In these landscapes, a boom in foundations of museums, archives, and libraries helped turn the nineteenth century into a long century of collection and preservation. It is not arbitrary, then, to count photography alongside the telegraph and the steam engine among the emblematic technologies of the era. Photography seemed to make it possible, for a moment, to arrest a greatly accelerated time by capturing it in an image.

Even if, in early photography, this moment ranged from several minutes to whole hours, that did not affect photography's promise to immobilize: It continued to reverberate in literature, painting, and sculpture. At the same time, it was echoed in contemporary science in a variety of ways. Various attempts were made to capture the speed of sound, electricity, and light. In addition, the cultural process of perception was held in suspension by a broad range of techniques, as Jonathan Crary has shown, including stereoscopes, panoramas, and carefully staged spectacles that required a sharp gaze and focused attention.[29]

In similar ways, Helmholtz's time experiments operated as *deceleration devices*—in part because the curves were authentic snapshots of physiological processes (the recorded contractions on average lasted for a total of half a second). Yet Helmholtz's experiments also had decelerating effects in that they required the experimenter to continuously observe the processes he induced and the results his repeated experiments produced. The comparison with photography seems particularly fitting here, since the materiality of Helmholtz's experiments defined a kind of frame: that is, they cut out or, as it were, cropped a specific part of reality in the lab.[30] By combining and contrasting instruments, recording devices, and human as well as nonhuman organisms, his experimental setups circumscribed a space that served to capture, trace, and measure psychophysiological processes characterized by high velocities.

Within this space, Helmholtz's experiments defined their own temporality, a specific kind of "laboratory time."[31] This form of temporality can be described as an *artificially created present* that is marked by the simultaneous presence (or availability) and functioning of all components of the experiment, here: frogs or human test subjects, energy sources, rotating recording surfaces, notebooks, pencils, and so on.

It goes without saying that in actual research practice, this presence was far from momentary. In fact, the present time of Helmholtz's studies always remained an unfinished, evolving one—not a single immobile shot but a succession of changing views. The present time characteristic of these investigations was, as already indicated, an open-ended process of repetitions and differences that concerned both the results, which were unexpected, and the precise object or topic of Helmholtz's research, which was not fixed once and for all. In other words, the experimental process went forward and backward, took ludicrous detours as well as unexpected shortcuts, and most of the time did not reach the goals initially targeted.[32]

Helmholtz's *theoretical* views appear quite remote from this notion of selective repetition. In Deleuzean terms, Helmholtz's theory of the "conservation of force" makes him seem like a prominent representative of a mechanistic version of the eternal return. His *practical* work as physiological experimenter, however, situates him differently, side by side with Kierkegaard, Nietzsche, and Péguy—authors whom Deleuze characterizes as the most innovative among the "great repeaters" of the nineteenth century.[33] Similar to the work of these three philosophers, the Königsberg experimenter's work aimed at "producing . . . a movement capable of affecting the mind outside of all representation."[34] And indeed, the research process Helmholtz's time experiments involved him in often appears as a radical encounter with the individual, as a drifting reproduction of the diverse.

In this process, problems undergo significant changes. They are shifted and displaced in unpredictable ways. Carefully assembled experimental setups are suddenly retooled or replaced by half-forgotten ones—to the point that we may well ask whether or not we are still dealing with the same experiment. Retrospectively, Helmholtz speaks about such laboratory experiences as "wanderings" (*Irrfahrten*) and compares himself

> with an Alpine climber, who, not knowing the way, ascends slowly and with toil and is often compelled to retrace his steps because his progress is stopped; sometimes by reasoning, and sometimes by accident, he hits upon traces of a fresh path, which again leads him a little farther; and finally, when he has reached the goal, he finds to his annoyance a royal road on which he might have ridden up if he had been clever enough to find the right starting point at the outset.[35]

The present study, accordingly, highlights the slowness and laboriousness of experimental progress. It emphasizes the accidental detours and surprising pathways created in and by the daily practice of laboratory research.

Despite our rejecting them earlier as anachronisms, the cuts and crossfadings that we will encounter along the way make it tempting to turn from photography to cinematography for a promising model for the process in question. The argument would be as follows. Helmholtz's time experiments do not just record movements, that is, muscle contractions. They are not just "cinematography" in the most literal sense. They also provide means for cutting and editing the recorded contractions in almost arbitrary ways, as if they were filmed.

This, precisely, is where the epistemic potential of his analytical experimentation lies. We are dealing with *sequences* that are essential for capturing, representing, and measuring the physiological processes in question. The content of these sequences—for example, movements of hand and mouth in a test person whose nerves were stimulated at various points on his or her skin—is relatively simple. However, frequent repetitions and differentiations were required to record the curves and to make the appropriate measurements.

This preliminary staging, the *mise en scène*, as it were, is another element that made Helmholtz's time experiments a relatively slow practice. Among other things, this practice involved laboriously fabricating and exploring instruments, preparing muscles and nerves, training test subjects, calibrating instruments, recording and copying curves, and writing down and calculating numerical results. It also involved handling disturbances that intervened from outside or inside of the experimental frame.

From this perspective, Helmholtz's research machines appear as material counterparts to the ethos of the "patient drudge" that scientists began to embrace in the second half of the nineteenth century, faced as they were with processes of cultural acceleration.[36] What we are dealing with, then, is not a fixed *Retard en verre* à la Duchamp, but a slowly developing delay in brass and steel, muscles and nerves, paper and electricity.[37]

Proust's experiment

Accomplishing decelerations with the means of acceleration: This could be the operation that makes the Helmholtz curves "epochal" images—in a double sense. On the one hand, they mark the beginning of a new epoch in the quantitative study of organic life. On the other hand, they visualize a suspension, a stopping of the seemingly continuous stream of psychophysiological life. In other words, they show something like an *epokhē* in the phenomenological sense.

Titles are an important element of all modern images. In the case of Helmholtz's images, however, the caption is truly remarkable: It reads "*temps perdu*," lost time. That is the term Helmholtz introduces for the most significant part of the recorded movement process, namely the short initial interval during which the muscle, after its stimulation, does not react at all. In the French context, the expression "temps perdu" was quickly taken up and widely used: in physiology manuals and physiological diction-aries, as well in the increasingly popular writings of Marey. Eventually, it would surface in the work of Marcel Proust, which was epoch-making in its own way.

As is well known, Proust kept in contact with the Paris medical and bio-logical scene. His father, Achille Adrien Proust, was a physician and epide-miologist who even collaborated with Marey for a while.[38] It is perhaps no surprise, then, that his son's literary works and translations reveal a striking familiarity with physiological laboratory techniques such as sphygmography and chronophotography. In addition, the initial title of Proust's famous novel, *A la recherche du temps perdu*, made use of scientific terms that were used by the graphical experts of Marey and his school. In the lab, "*Les inter-mittences du Cœur [The Intermittencies of the Heart]*" referred to short and unpredictable interruptions of cardiac activity.[39]

If we accept this lineage of the expression "temps perdu," further con-vergences between the searches for time conducted by Helmholtz and Proust come within our grasp: the importance of photography as a form, the criticism of the performance of human sense organs, and a related proj-ect defined against this background, namely the project of building a kind of telescope for the accurate perception of time. The work done by Helm-holtz and Proust also converges in the careful isolation of time experimenters and test subjects from the outside world; the role of journals, telegrams, and other means for rapid communication; and, eventually, the relation of signs to truth and, therefore, of truth to time.[40]

Yet Helmholtz's research neither started with the phenomenon of lost time nor stopped there. More often than of "*temps perdu*"—or, in German, "*verlorene Zeit*"—he spoke of "*Zwischenzeit*," meaning "interim" as well as "intervening period." Helmholtz initially used this expression to describe the delay phenomenon in muscle contraction. Subsequently, he used it as a kind of guide in his continuing search for breaks and discontinuities in the process of psychophysiological life. As "Zwischenzeit" was (and still is) a common term in German, comparable to the English term "meantime,"

Helmholtz here makes productive use of everyday language on the level of scientific discourse.

At the same time, however, Helmholtz's phrasing alludes to a more specific use of the term. In the theater, "Zwischenzeit" designated the time that elapses between two acts of a drama, the intermission during which the audience does not see any action taking place on stage. Translated back into Helmholtz's experiments, "Zwischenzeit" would then refer to the fact that the frog muscle, immediately after its stimulation, hesitates before recording any curve. Does this also mean, however, that nothing happens at all during this fraction of a second? In other words, is "Zwischenzeit" really a lost time, an empty time, devoid of all events?[41]

Besides theater, there is another possible context for the term. Helmholtz's adoption and use of "interim" might also be reminiscent of Johann Wolfgang von Goethe's book on color theory, with which he was quite familiar. In fact, in the historical parts of Goethe's *Zur Farbenlehre* (Color Theory), "Zwischenzeit" functioned as the name for the "gap" (*Lücke*), in the history of color theory, between antiquity and the Renaissance. In this context, Goethe speaks about those "quiet dark times" that are "the incalculable, incommensurable element of universal history."[42]

Helmholtz was well aware of the fact that the physiological discontinuity between stimulation and contraction could not simply be calculated. However, measurements of this intermediary time were possible, as his experiments had shown in rather impressive ways. These measurements even seemed desirable, since, like Goethe's universal history, they pinpointed a kind of time in and through which the human being appeared as "unknown to himself"—a being full of possibilities and dangers. As we will see, experiments on human beings were indeed a focus of Helmholtz's time research. Parallel to his work with frog muscles, he conducted what he concisely called "human time measurements" (*Menschenzeitmessungen*).[43]

As a result, however, his experiments exposed Helmholtz to an additional problem of representation. The question was no longer just how to transpose complex phenomena and austere numbers into persuasive curves. The problem was also how to determine the place and status in human experience of the results of physiological research. At this point, the fact of nervous and—and in the case of human beings—cerebral intervals acquires an almost poetic character. Helmholtz comes very close here to playing a role later performed by Proust, that of the avant-garde writer who, in order to reach his audience, has to provoke his readers and even ask too much of them.

There was little doubt from the physiologist's perspective that the propagation of stimulations through the nerves and the brain takes place at a relatively low speed. However, it was also obvious to Helmholtz that this state of affairs was accessible neither to common sense nor to introspection or recollection. At one point in the course of his experiments, he even has to admit: "We have never perhaps experienced anything similar."[44] And yet Helmholtz insisted on the fundamental role interims play in human experience and behavior. If they did not exist in the short form he had determined with great precision, "our self-consciousness would lag far behind the present."[45] To this day, this splitting of the subject based on laboratory time proves to be as irritating as it is interesting.

Helmholtz and the "graphical method"

The graphical method constitutes a significant chapter in the history of the modern life sciences. Ever since Sigfried Giedion's suggestive juxtaposition of physiological and artistic movement studies—of the Weber brothers and Marey on one side, Marcel Duchamp and Wassily Kandinsky on the other—historians of science and art have repeatedly discussed the traces, notations, and writings laboratory physiologists produced in the late nineteenth century to investigate, with gradually increasing accuracy, the vital functions of organic individuals and to advance the processes by means of which "mechanization takes command."[46]

If the present study goes back to the history of these traces, it does so not to highlight their baroque aesthetic one more time. It does so to show that the history of these curve drawings is an integral part of a history of modernism that ascribes central importance to the problem of social, cultural, and technical synchronization. For sociologist Norbert Elias, the drive to increase synchronicity is indeed one of the most striking features of modernity: "With growing urbanization and commercialization, the problem of synchronizing the growing number of human activities and of having a smooth-running continuous time-grid as a common frame of reference for all human activities became more urgent."[47] The following investigation will refer to such requirements of social organization through the thick, as it were, of historical documents and instruments. It also will link up with recent studies in the history of the graphical method, studies presented as explicit contributions to the history of culture and media.[48] Nonetheless, it differs from these works in that it targets one specific example: Two images made in 1851 that show a total of six curves.

The focus on such a concrete example makes a methodological argument. The present study is not just a contribution to the cultural and social history of science objects, or "epistemic things," as Hans-Jörg Rheinberger puts it. It also aims at contributing to a historical "epistemology of the detail."[49] It takes as its starting point *one thing* and attempts to follow this thing, in a Latourian sense, into the spaces of its "construction" as well as into the time of its "translation," or dissemination. To attain this goal, this book works, at least in part, with high resolutions. It generates and accepts a wealth of detail(s) in its technical, pictorial, and textual sources and their respective translations, transpositions, and transfers that might come as a surprise to at least some of its readers.

Generally, we have no problem with closely "reading" artistic and literary works over and over, especially modern classics like Kafka or Duchamp. Only sporadically, however, have scientific texts and images received similar attention, and those that did were typically major treatises or seminal books, such as *On the Origin of Species*, *Kosmos*, or *The Interpretation of Dreams*. In what follows, by contrast, scientific reports, articles, and essays take center stage. The present study is also intended to be a plea for devoting more attention, patience, and historical expertise to science's short texts and small images.

Given the importance of Helmholtz's time experiments in the history of the life sciences, however, it is not surprising that there is a variety of studies—from Edwin G. Boring to Hebbel E. Hoff and Leslie A. Geddes, from Wilhelm Blasius to Richard Kremer and Timothy Lenoir, from Kathryn Olesko and Frederic L. Holmes to Robert Brain, M. Norton Wise, and Soraya de Chadarevian, as well as from Claude Debru to Stanley Finger and

Nicholas Wade.[50] As are Holmes and Olesko, I am above all interested here in reconstructing the "investigative pathway" that led from Helmholtz's initial interest in muscle activity to his study of nervous processes and from curve recordings to precision time measurements—and back again.

In addition to the relevant (and well-known) articles published by Helmholtz, my reconstruction makes use of a new corpus of laboratory notes, manuscripts, and images. For example, taking into account the recently discovered curve images makes it possible for the first time to fully understand how these epochal drawings were traced, transported, and eventually transposed into published figures. I will also show how much the curves depended on Helmholtz's interpretations and interventions. The curves were meant to be self-evident, but much explanatory work was required for Helmholtz's audience to actually see what the curves were meant to show. Finally, it becomes clear that theses images owe their existence to the difficulties Helmholtz encountered in communicating his experiments in the initial report. The numbers he had given there met with considerable suspicion—even among the physiologically informed public.

Another finding of this book concerns the importance of Helmholtz's "human time measurements." Even these measurements inscribe themselves in the more general intention of concretizing the relatively abstract results of measurements in nerve-muscle samples taken from frogs. Moreover, Helmholtz's "human subject experiments" (*Menschenversuche*)[51] are an event of crucial importance for later research at the interface between physiology and psychology. It is true that statements on his time experiments on human subjects published by Helmholtz are rare. However, as Klaus Klauß has shown thanks to an unpublished report to the Berlin Physical Society, Helmholtz himself attributed great significance to these measurements in the early 1850s. In fact, Helmholtz's human time measurements were later adopted, replicated, and differentiated by numerous experimenters in the fields of neurobiology, brain research, and the cognitive sciences—among others, Adolphe Hirsch, Rudolf Schelske, Franciscus Donders, Sigmund Exner, and Wilhelm Wundt, the often-quoted "founding father" of experimental psychology.[52]

Finally, Helmholtz's laboratory logbook from 1850 allows for redefining the scientific role of Olga Helmholtz in striking new ways. While Holmes and Olesko mention this logbook, they do not draw on it in their analysis. They only highlight the fact that Helmholtz's wife Olga functioned in the time experiments as an "assistant" (*Gehülfe*). What I show here is that her role included writing clean copies of Helmholtz's texts prior to publication, reading off instruments, and recording data, as well as taking notes at the laboratory bench. In the context of Helmholtz's human time measurements, Olga even acted as a test subject and, in turn, conducted similar experiments without her husband's supervision.[53]

Like the works by Robert Brain and Norton Wise, the present book puts Helmholtz's experimental work into cultural and social contexts to deflate the still-popular image of a scientific genius largely working in splendid isolation. With this goal in mind, Wise, in his recent studies, has related Helmholtz's curve method to the history of art and technology in Wilhelminian Berlin.[54] However, similar contextualizations are still lacking for the Pouillet method, which played a crucial role in Helmholtz's time experiments.[55]

With the present study I attempt to fill this gap, while also aiming at developing a more precise notion of the culture *and* technology folded into Helmholtz's time experiments. In this context, I will highlight the fact that Helmholtz's concrete point of reference, the Pouillet method, was largely rooted in a merging of electric telegraphy with mechanical clocks that by itself had far-reaching consequences for the history of time and synchronization in the nineteenth century.[56]

"Context" in this connection does not refer only to Berlin. In contrast to previous investigations, this book situates Helmholtz's experiments in a cultural and technological environment that is much more expansive. Besides Berlin and Königsberg, Paris and its Academy of Science are among the principal sites for investigating Helmholtz's time experiments. Accordingly, historical actors enter the picture whose crucial importance for these experiments has rarely been acknowledged.

Such is the case of the Berlin-based electrophysiologist Emil du Bois-Reymond. Although he is often quoted as Helmholtz's older friend and correspondent, his active role in interpreting, translating, and disseminating Helmholtz's reports and articles, as well as his attempts to demonstrate practically Helmholtz's time experiments, have not been adequately appreciated. Du Bois-Reymond was born into a Swiss-Huguenot family. His father, a trained watchmaker, worked in the Prussian Interior Ministry until 1848. Emil du Bois-Reymond spoke excellent French and from an early age was personally acquainted with members of Berlin's scientific and political elite. As a result, he was a perfect fit for the role of translator and mediator for Helmholtz's physiological research.

Another neglected actor is the Italian physicist, telegraph expert, and pioneer in electrophysiology, Carlo Matteucci. Following Hoff and Geddes, I will argue that in his experimental studies on the physiology of muscle contractions, Matteucci applied the graphical method *and* an electromagnetic timing device two years before Helmholtz did.[57]

On a more general level, this study transforms into an internationalized narrative a topic that until now has largely appeared as a chapter in the history of German physiology, an account in which (as Soraya de Chadarevian and Gabriel Finkelstein are right to emphasize)[58] Franco-German relations are of particular importance. It has been said that with regard to the history of twentieth-century philosophy, France and Germany have one and the same indivisible history. In this book I show that to some extent this holds true for the history of nineteenth-century physiology as well.

My final point concerns the relation between thing and medium. Most historical studies on Helmholtz's experiments have taken "time" to be the *object* of this research. No account so far has understood time to also be a *means* of these experiments. In the case of Helmholtz, as in many other fields of science studies, a largely spatialized view has been dominant. I, too, am interested in reconstructing the experimental assemblages and networks in which he carried out his research and published his results. However, when in this connection I use the term "research machines," which I do repeatedly, I want to emphasize the temporal aspects of scientific practice. On the one hand, "machine" makes reference to the discursive practice of historical actors. For example, Helmholtz talks about his "frog drawing machine" and almost tenderly calls one of his early experimental setups "my little machine."[59] On the other hand, however, "research machine" highlights the cultural, historical, and analytical potentials of a term only occasionally used in science studies to establish connections between laboratory practices and the history of technology and culture.[60]

Like "network" and "system," "machine" stands for the materiality and situatedness of scientific practice. However, unlike other terms, it also underscores the temporality and dynamics of laboratory action. As a result, time in this book figures not only as the notoriously scarce resource of the individual scientist but also as a characteristic of great importance in the performance of technical equipment in the lab. In addition, however, time will turn out to be a crucial factor in organizing how research machines function in the laboratory as well as in communicating and disseminating experimental results. Helmholtz often complained about the difficulties that arose because he had to simultaneously control a variety of instruments in his lab. By publishing a preliminary report about his findings, he tried to secure his scientific priority—which essentially is a temporal issue. Furthermore, his turn to the method of the curves was motivated by the speed in demonstrating physiological facts this method made possible. In the notion of "research machines," all these aspects of time reverberate.

By highlighting these dimensions of temporality, I aim in this book to render the peculiar density of scientific time more palpable. Only in very rare cases does the time of science reduce itself to a uniform chronology, a smooth course of action without any complications or detours. To quote author and curve artist Laurence Sterne, science does not progress along a

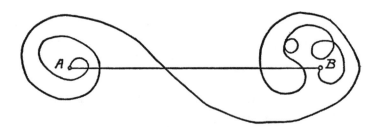

FIGURE 3. The difference between "dogmatic" and "historical" accounts of scientific developments. According to George Sarton, there are always at least two roads from one scientific discovery (A) to the next (B), "the long 'historical' one which leads to the discovery of B, and the 'dogmatic' one which leads from A to B in the simplest and quickest manner." Reproduced from George Sarton, *A Guide to the History of Science: With an Introductory Essay on Science and Tradition* (New York: Ronald Press, 1952), 39. © 1952 George Sarton.

straight line drawn "by a writing-master's ruler" between A and B. Instead, it takes place in curves, with "transgressions" that are hardly predictable, including "the common *ins* and *outs*."[61]

In this sense, then, this narrative highlights the difference between a "dogmatic" and a "historical" account of the process of science, as others have done (Figure 3). In addition, however, I seek to move the focus from physiological time to the time of physiology, from an object to a means of research. The line that traces this move is the one we are interested in here.

ONE

Curves Regained

> The element of time, in general, is discernable in the line to a much
> greater extent than it was in the case of the point: length is a concept
> of time.
>
> —WASSILY KANDINSKY[1]

On September 1, 1851, a note from Hermann Helmholtz was read in the
Académie des sciences in Paris. Helmholtz had been professor of physiology and anatomy at Königsberg University since 1849, and in this note,
he reported on his continued studies on the propagation speed of nerve
stimulations. A routine event in nineteenth-century scientific circles, we
might think.

On closer inspection, however, we quickly find ourselves caught up in a
deconstructive movement that unstoppably slides from author to writing
and from text to technics. We might accept that the title of the note (which,
in translation, reads "Second Note on the Propagation Speed of the Nervous Agent") has little to do with its content.[2] But how are we to deal with
the fact that it wasn't Helmholtz who wrote the French text submitted to
the Academy of Sciences? What are we to make of the fact, borne out by a
comparison of manuscript and print version, that some parts of his text

were not printed? Was Helmholtz consulted, or did the Academy act on its own initiative? And finally, what are we to think of the fact that the precision of the images, which Helmholtz discusses at length in the note, becomes intelligible only if we understand the details of the drawing machine, which, however, is given only a rough description?

This state of affairs is further complicated by the fact that Helmholtz's note of September 1, 1851—the title says it clearly—was not his *first* on the topic of the propagation speed of nerve stimulations. The Königsberg physiologist had been busy with research on the topic for at least two years. Preliminary results were communicated to the Paris Academy of Sciences at the beginning of 1850. By using Pouillet's method and adjusting his research machine accordingly, he had by that time precisely determined the speed at which a stimulation travels through a nerve—or had managed at least to define a precise range of speeds. In the motor nerves of a frog, the speeds he measured were in the range between 25 and 43 meters per second.[3] These values were significantly lower than the values known at the time for the speed of light, of electricity, or of sound.

For a long time, physicians and scientists held that nervous action might be due to an "immaterial" or "imponderable principle" and thus might take place with inconceivable rapidity. Helmholtz's time measurements made clear that the propagation of the stimulation had to be a materially based process, in which the inner components of the nerve were set in motion. If that were not so, the relative slowness of nervous action would be very difficult to account for.

Even before publication in France, the news had spread, especially in Helmholtz's hometown, Berlin. There, even Alexander von Humboldt was impressed: "So curious a discovery speaks through the astonishment it provokes,"[4] he wrote to the young physiologist in Königsberg—and insisted on conveying Helmholtz's first message (the "Note") to the Academy in Paris in January of 1850.

Yet Helmholtz hadn't quite been out to provoke astonishment. He was concerned with having a new scientific fact, an achievement, acknowledged to be such; he was concerned with precision and rationality. In further measurement experiments he therefore sought to render even more precise his determination of nerve time. The result he finally obtained, after months of experiments, was an average value of 26.4 meters per second. This new result, which again was based on work with prepared frog samples and the application of Pouillet's method, had already been the subject of a detailed, ninety-page publication in German a year earlier, in the fall of 1850.[5] Now he communicated this result, in shorter form, to the Paris Academy as well.

The real purpose of his message, however, was a different one. The "Second Note" in fact mentions the new measurement only incidentally. It focuses on presenting a "new method" meant to facilitate time research on muscles and nerves of cold- as well as warm-blooded organisms—of frogs, for example, and of human beings.[6] This new method was no longer based on electromagnetism, as the Pouillet method had been, but on largely mechanical means: the self-recording of muscle movements in the form of curves.

These curves were scratched (with a stylus attached to a hanging frog muscle) into a layer of soot (applied with the help of an open flame, such as a candle) on a rotating cylinder. Helmholtz praised the advantages of this new method, especially that it reduced the number of required repetitions of the experiment and that it dispensed with the elaborate calculations necessary in the Pouillet procedure.[7] Ultimately, however, these were not the goals he sought to achieve.

The decisive advantage of the curve method was instead that it made his already completed time measurements more plausible. Using the Pouillet method meant painstakingly calculating the propagation speed of the nerve stimulation from individual values noted in table format. This resulted in a naked number standing at the end, an average value. The curves, by contrast, made movements and, through these, time visible. They pictured time in a lasting trace. They thus allowed other scientists, both physiologists and nonphysiologists, to get a faster and better overview of the phenomenon in question.

This, at least, is the claim made by the text of the "Second Note"[8]—a claim that comes up short in a remarkable way: there is no figure that would back it up. And for a simple reason: The periodical in which the Paris Academy published its weekly Monday sessions, the *Comptes rendus hebdomadaires des séances de l'Académie des sciences*, was not laid out to include reproductions of illustrations. The *Comptes rendus* (or "proceedings") were chronicles, written records, in the literal sense: Their semiotic world consisted solely of the alphabet and of mathematical expressions and arguments. There was no space for other forms of chronicling, of writing time.

It was not until thirty years later that an exception to this rule was granted, an exception that, significantly, also concerned physiology and time. In 1882, no less a figure than Etienne-Jules Marey, famous for his innovations in graphical methodology, was given permission to reproduce in the *Comptes rendus* chronophotographs he had taken of the walking movements of a man and of a horse.[9]

At the beginning of the 1850s, such reproductions would have been very difficult to put into print, if only for technical reasons. Helmholtz would have been aware of this, especially since this was not the first time he published in the *Comptes rendus*. He nonetheless included images with the manuscript he submitted, probably to convince those members of the Academy attending the session of his new method's visual potential. It is unclear, however, whether the Academy members assembled on September 1, 1851, did in fact look at the Helmholtz curves and whether the explanatory text specifically composed for the occasion was read aloud.

Be that as it may, a week later, during the session of September 8, a commission was set up to examine Helmholtz's two communications more closely. This was one of the many routines by which the Paris Academy could react to the announcement of scientific news. After a first review had released the text for publication in the *Comptes rendus* (which, in the case of nonmembers like young Helmholtz, was a special honor in itself), a detailed assessment would evaluate the conclusions and claims of the message in question. Besides François Magendie and Pierre Flourens, two of the most eminent French physiologists, the committee also included the physicist Claude Pouillet, the namesake of the procedure for measuring time that Helmholtz had initially used.

But the commission was never to publish any report. Apparently, Magendie was very busy anyway; not just within the Academy, where he sat on a number of other committees, but also outside—especially with advancing the career of his former assistant, Claude Bernard. In 1848, Bernard had finally come forward with discoveries of his own. Magendie was firmly convinced of the potential of this work to provide new impulses for physiology in France, hence his enthusiastic support.[10]

Another member of the commission, Pouillet, was also reticent, possibly because the method he had developed and that Helmholtz had put to such productive use initially was criticized in the "Second Note" as too complicated and elaborate.[11]

This left Flourens. And indeed, he was the only one to take action. At least he acknowledged receipt of the manuscript of the "Second Note" by scribbling the word *vu* (seen) on it—which does not necessarily mean that he had carefully read the text or taken a good look at the images included.[12]

Shortly after the note had appeared in print, both the text and the images went into a file. And that's where they are still today. They have in the meantime been stamped and moved into the archives of the Académie on Quai de Conti, in the midst of innumerable other files, folders, and documents.

No one knows if the curves have been seen by a human eye since. There is no checkout card on which possible users of the document would have entered their names. There is no record of earlier references to these curves, never mind of earlier reproductions, at least not to the knowledge of Helmholtz experts. It thus seems that for more than one hundred and fifty years, these undulating forms were engaged in a dialog with the waves of the Seine, a dialog as mute as it was blind. The only thing that can be said with certainty is that they are the only curve drawings currently known that indexically refer to Helmholtz's time experiments (see Figures 1 and 2).

There is no other record—at least not on paper—testifying to these experiments. That is, there are no other *things* left of them. All instruments and contraptions, all cylinders, styluses, and contact parts Helmholtz made use of to make his measurements and draw the curves in Königsberg, have been lost. They were lost as early as ten years later, when Helmholtz, after some time in Bonn, joined the University of Heidelberg.[13]

It is also not clear who really wrote the "Second Note." The text published in the *Comptes rendus* seems to leave no space for doubt about the authorship of "M. H. Helmholtz (de Königsberg)." But a look at the manuscripts changes the picture. To the extent that it is possible to abstract from the difference between the old German and Latin scripts, there seem to be hardly any similarities between the manuscript preserved in the Paris Academy archives and other Helmholtz manuscripts. Perhaps Olga Helmholtz, his wife, was responsible for writing down this text. We know at least that around the same time she produced clean copies of other texts of her husband's prior to publication.

At any rate, the draft of the "Second Note" preserved among Helmholtz's papers in Berlin is not written in Helmholtz's hand.[14] As had already been the case for the first "Note" to the Paris Academy in January of 1850, it was Helmholtz's friend and colleague du Bois-Reymond who had authored the French text. "To author" seems indeed to be the right verb. Du Bois-Reymond had not only translated the first note but also thoroughly redacted it. The case of the "Second Note" is all the more complicated for the absence of a German source text. There is no version extant in either manuscript or printed form.[15] It is thus not possible to determine the extent to which du Bois-Reymond cowrote the "Second Note" to the Academy.

For these reasons, we are moving through a real thicket of writings, both human and nonhuman: On the one hand, there are the numbers and letters put on paper by male and female agents; on the other, there are the straight and rolling lines drawn by frog muscles, levers, and styluses on a rotating cylinder. Only the one who ultimately figures as the author seems nowhere to be found in this thicket: Monsieur Helmholtz.

A careful reading of the four printed pages of the "Second Note" only increases the confusion. Almost every expectation today's reader might bring to such a text is frustrated. The title of the publication announces information about the propagation speed of the nervous agent. But the first part, at least, does not keep the title's promise at all. Instead of a report about measuring the propagation speed of stimuli in *nerves* we find a summary of time-measurement studies of the activity of *muscles*. In addition, the only results communicated here had already been published in German. The first part of the "Second Note" relies entirely on the ninety-page treatise Helmholtz had published one year earlier in the *Archiv für Anatomie, Physiologie und wissenschaftliche Medicin*, which contained the first detailed report on his time experiments.[16]

There is, therefore, nothing that is obviously new, and yet the very first paragraphs of the "Second Note" are remarkable. In the description of the individual phases of muscle contractions, they introduce a terminology not present in the ninety-page treatise. This terminology was to be widely publicized, at least among Francophone scientists, thanks to its inclusion in the physiology textbooks of Maurice Arthus and Henri Beaunis, in Charles Richet's physiological dictionary, and in books of Marey's such as *Du mouvement dans les fonctions de la vie* (1868), *La machine animale* (1873), and *La méthode graphique* (1878).[17]

The "Second Note" distinguishes among three phases of contraction. The first is the short delay between the sudden stimulus and the beginning of the contraction. Helmholtz (or du Bois-Reymond?) calls this interim *temps perdu*, a term emphasized by underlining when it is used for the first time (Figure 4).[18] In frogs, this "lost time" amounts to 0.01 seconds on average. The second phase is the contraction proper, in which "the tension of the muscle increases." Its average duration is 0.08 seconds. Finally, the third and longest phase, "the decline in the tension of the muscle until its complete relaxation," lasts between 0.30 and 1.00 seconds.[19]

Only then does the text turn to its topic proper, the propagation speed of the stimulation in the nerve. But it does so indirectly. First there is a summary of the discussions on the topic already contained in the 1850 German ninety-page treatise. Then there is a discussion of Pouillet's

FIGURE 4. Draft of Helmholtz's "Deuxième note," written by his friend and colleague Emil du Bois-Reymond (ca. 1851). When introduced, the term *temps perdu* is marked by underlining. It refers to the first phase of the contraction recorded, the minimal delay between the electrical stimulation of the muscle and its contraction. Reprinted with permission from Archiv der Berlin-Brandenburgischen Akademie der Wissenschaften, Berlin, Helmholtz Papers, NL Helmholtz, 526.

method, on which all divisions between phases and measurements of time discussed up to that point had been based. This method, we now read, is an "ingenious procedure"; its application in physiology, however, entailed "great difficulties."[20] Above all, securing results against all doubt necessitated "long and tiresome series of observations." The times Pouillet's method could measure with great accuracy may have been extremely short. Doing the measurements, however, was time-intensive business.

At this point, the scientist-author's first-person voice enters forcefully:

> I therefore looked for a more expedient method, and I have the good fortune to have found one whose principle is very simple and that is much more easily put into practice. It allows us to show the same facts in the space of a few minutes, with only a small number of experiments.[21]

Helmholtz's turn to the curve method was, on this reading, due to an economy of time concerning the labor time of the experimenting researcher on the one hand but also, on the other, the time of demonstration: that is, the time needed for showing a scientific fact. "A few minutes"—compared to the elaborate measurements required by Pouillet's method or to the effort of reading a ninety-page treatise in German, that would really constitute a reduction in the time spent.

Yet this is not about showing, as such. The "Second Note" is clear on this point. The concern is "to show the same facts" (*démontrer les mêmes faits*). This sheds more light on this second note's character as a whole. By this point, at the very latest, any expectation of new scientific facts will have been frustrated. Obviously, what is presented is an alternative method for demonstrating physiological facts that, as such, are already known. We are thus dealing less with a communication about innovative science than with a message about a new technique (if we may for once insist on this separation between the technical and the scientific).

There is no doubt the Académie would have been the right place for such an announcement. In 1839, this had been where the daguerreotype was introduced, and as time went on, a large number of telegraphs, precision chronographs, and other novel technical objects were presented in this forum. Be that as it may, the "Second Note" contains a first blending of repetition and difference, an accelerated Once Again for an experiment already performed but furnished with different instruments—and with a different goal. The focus is no longer on measuring but on showing.

A new image of life

The modern life sciences' struggle for recognition and prestige has always been also a struggle about images. This is as apparent in today's brain sciences and genetics as it was in nineteenth-century embryology and bacteriology. In recent years, historians of science and of art have time and again demonstrated how much Darwinism is based on thinking and speaking in images. Research in the history of experimental physiology soon pointed in a similar direction. Numerous studies on Marey and the "graphical method" have shown clearly how much the laboratory revolution of the life sciences in the nineteenth century was not only a scientific but also, in Nicholas Jardine's phrase, an "aesthetic accomplishment."[22]

This accomplishment, however, was not limited to the comparatively narrow spaces of physiological research. Instead, it quickly expanded to university teaching and the nonacademic public. In fact, the aesthetics of the laboratory revolution brought about by physiology is particularly evident in the lecture halls of the institutes founded at the time. Since the late 1860s, physiologists such as Jan E. Purkyně and Johann N. Czermak set out to fashion and publicize a new image of life by animating (i.e., endowing with life) drawings as well as by advancing shadow play, projections, phenakistoscopes, and other optical media. They projected greatly magnified outlines of twitching frog hearts on screens in darkened lecture halls or set up magic lanterns and movable glass panels in such a way that recordings of curves could take place before a large audience almost in real time (Figures 5 and 6). At the beginning of the 1870s, Czermak, with his own money, built a *Spektatorium* in Leipzig. Its main purpose was to present "in detail to listener's immediate visual perception" the "inaccessible and strange processes" physiological science sought to recognize and to explain.[23]

Some years later, Helmholtz's Berlin friend and colleague du Bois-Reymond was also to emphasize the role of the visual in communicating

FIGURE 5. Projection of a beating frog heart in the lecture hall of Johann N. Czermak's private laboratory in Leipzig (1872). Reproduced from Anonymous, "Czermak's physiologisches Privatlaboratorium und Amphitheater in Leipzig," *Illustrirte Zeitung*, no. 1556 (1873): 305–7, here 305.

physiological facts. As he explains at the opening ceremony of his institute on the corner of Wilhelmstraße and Dorotheenstraße, "seeing for oneself and ascertaining for oneself" is the physiologist's most important task, even if "at first" it takes "some effort."[24] For that very reason it is desirable, he continues, to "demonstrate the phenomena in as evident a manner as possible" during lectures. Without physiological demonstration, every oral presentation would remain "sterile."[25]

There aren't any remarks on the subject by Helmholtz, in his later years, that would be nearly as explicit on the role of seeing and showing in physiology. Yet he had already introduced such techniques in his teaching at Königsberg University by the 1850s. In 1852, for example, he tells du Bois-Reymond that he had demonstrated the latter's electrophysiological experiments "very palpably" before a group of twenty students by means of a mirror galvanometer and the corresponding projections of (sun) light.[26]

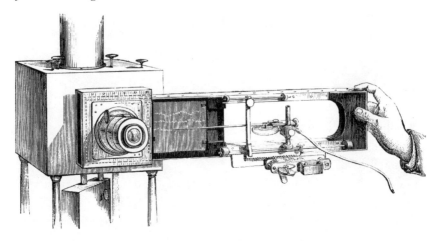

FIGURE 6. Projection polygraph for use in lecture halls (1868). To present a "new image of life," Marey used, among others, devices that could project curve recordings—of a heart beating, for example—almost in real time. The curves were scratched, at distances that could reach more than forty feet, onto a glass plate covered in soot that was fed through a magic lantern and projected an animated image on the wall. Reproduced from Etienne-Jules Marey, *Du mouvement dans les fonctions de la vie* (Paris: Baillière, 1868), 190.

Two decades later, this use of light beams as "weightless sensing lever" had become a standard element in the object lessons of Czermak, du Bois-Reymond, and many other physiologists. Remarkably, however, this significance of images is already apparent in the "Second Note." Experimental physiology's quest for "viewability," or *Schaubarkeit*, as Czermak put it, thus had already begun at a time when the discipline, in its modern form, was still emerging.

The technique Helmholtz had developed to this end for his time experiments is described succinctly in the "Second Note":

> Picture a smooth glass cylinder turning on its vertical axis at a uniform speed imparted to it by a clockwork with a cone pendulum. The cylinder's lateral surface is coated with soot. Facing the surface, at a very short distance, is a steel stylus, which nonetheless does not touch the cylinder unless the experimenter wishes it to. This stylus is capable of vertical movement; by way of a

system of levers, it communicates with the Achilles tendon of a frog's *musculus gastrocnemius* conveniently suspended near the cylinder. It is easy to see that the muscle, in contracting and lifting the stylus, can trace curves on the cylinder, curves whose horizontal coordinates are proportional to time and whose vertical coordinates are proportional to the shortening of the muscle.[27]

The illustration conjured by this succinct description corresponds to the simplified depictions of Helmholtz's myograph that were common in the 1860s and 1870s.[28]

We find such a figure in Marey's 1868 treatise on movement in the vital functions (Figure 7).[29] As with many other images of physiological experimental setups from that time, however, what is missing here (and from the text of the "Second Note") is a depiction of the energy sources and of the wiring of the various parts of the setup. As for the elaborate support frame in which the frog muscle was hanging on the machine that Helmholtz actually used in Königsberg, the humidified chamber in which the muscle was carefully housed, the nerve exposed at the tip of the muscles, and the electrodes affixed to the nerve at several points—all these are mentioned neither in Marey's figure nor in the description in the "Second Note." The same goes for the fact that the glass cylinder that Helmholtz had ordered from the Königsberg instrument-maker Egbert Rekoss was a piece carefully cut off a "cylindrical champagne glass."[30] .

There is just one more explanatory hint in the "Second Note." Helmholtz refers to a model for the "new method." Yet it is not Carl Ludwig, cited as the one who introduced the use of the kymograph in experimental physiology in 1847, nor is it James Watt, whose indicator diagrams are the oft-invoked model for Ludwig's kymographs. The reference, instead, is to the English physician, physicist, and rival of Champollion, Thomas Young. At the beginning of the nineteenth century, Young, in his *Lectures on Natural Philosophy*, had described a recording device in which a fixed "pin" traces lines on a quickly rotating drum. Helmholtz's reference to Young seems plausible insofar as Young, unlike Watt and Ludwig, explicitly presented his device as an instrument for measuring time, as "a *chronometer for measuring* minute portions of time."[31] Young, however, was not interested in its application in medical and biological research.[32] This goes some way toward

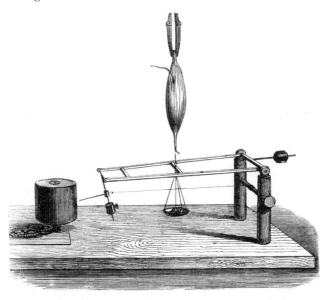

FIGURE 7. Marey's depiction of the myograph according to Helmholtz (1868). In the center, a muscle sample from a frog, held by a pair of tongs. From the left, a wire enters laterally into the muscle for the purpose of electric stimulation. Attached to the bottom of the sample are a metal frame and a weight for extending the muscle. The metal frame is fixed on axis on its right side, allowing it to move upward and downward. On its left side, a steel tip is affixed, ready to record muscle movements on the rotating and soot-covered brass cylinder (far left). In comparison with the device actually used by Helmholtz, however, this instrument is very much simplified (see also fig. 29). Reproduced from Etienne-Jules Marey, *Du mouvement dans les fonctions de la vie* (Paris: Baillière, 1868), 224.

explaining what Helmholtz considered the novelty of the method he presented. It did not consist in an original invention, as if ex nihilo, but rather in the unexpected yet precisely calculated joining of what had so far remained distant, in the creative assembling of the disparate.

A description of a few sentences' length and the mention of a name: At this point, those listening at the Paris Academy would probably have accepted the viability of Helmholtz's method as a method for recording muscle movements. What is likely to have remained largely unclear to them, however, was how one was to get from recording muscle movement to measuring nerve time. That required another step, which once again consisted in a repetition. The curve method repeated, with an alternative experimental setup, the experiments undertaken according to the Pouillet method in order "to show the same facts" once more. But the curves, too, had to be repeated now. They were written one above the other or, more precisely, the drawing stylus was made to go through a curve trace already scratched into the soot.

To bring about this second repetition, a "protruding cog"[33] was affixed to the flywheel below the rotating cylinder. This cog moved past a small lever that triggered an electric shock, making the muscle contract once. The rotating cylinder itself thus took care that the contraction curve would be recorded, and that it would be recorded at the same point on the cylinder. If all parameters coincided, the recorded curves coincided: "As long as nothing is changed in the setup of the experiment and the muscle is fully vigorous, all successively traced curves coincide exactly." This congruence of the recorded muscle movements was a necessary precondition for the depiction of nerve time—but it did not itself constitute this depiction.[34]

To that end, a first difference was introduced into the second repetition. Changing the location at which the nerve that triggered the contraction was stimulated led to a new curve whose "form" was "absolutely congruent" with the curve recorded earlier with the same muscle. The great difference, however, was that "the new curve [was] laterally displaced in relation to the other curve."[35] Between the two recorded curves, an interstice opened up, and this interstice derived from a difference in space and time: The stimulation had to travel different distances from the different points on the nerve, thus also needed more or less time to trigger the contraction.

Everything depended on the assumed identity of the repetition. Between the recordings of the two curves only one thing was allowed to change: the point at which the nerve was stimulated. All other variables were assumed unchanged, they were to be kept the same and thus also to be conserved.

The author of the "Second Note" was well aware that this constituted the Achilles' heel of an experimental setup in which the frog's Achilles tendon played an important role. Hence his remarkably defensive summary regarding the significance of the recorded double curves:

> I do not believe, accordingly, that it is possible to reasonably attribute the lateral displacement of the curve . . . to anything other than the longer trajectory the nervous agent is compelled to run through to operate the contraction of the muscle.[36]

That's it, astonishingly. That is the main argument put forward in the "Second Note"—but it is also exactly what had been announced in the text: the description of a method for showing a physiological fact that had been ascertained previously by other means.

This fact can be translated back into words as follows: Within a nerve, a stimulus needs time to get from one point to another. Ultimately, this had already been the content of Helmholtz's first message to the Paris Academy of Sciences. And there are no new results reported in the "Second Note" that would have been obtained from time measurements made according to the new method. The final sentence simply states: "Needless to say, moreover, that the measurements of the speed of the nervous agent obtained by these new means are in perfect accordance with those provided by Monsieur Pouillet's method."[37]

The attending members of the Academy were surely able to explain to themselves how the curve drawings could be translated into measurements of time—by taking into account the rotational speed of the glass cylinder, its circumference, and the distance between the points on the nerve. But there were no concrete data for this. It also remained unclear, in the end, how and to what degree of precision the "perfect accordance" of the old and the new measurements had been ascertained. The presentation of a new method, of an innovative technique, took center stage in the "Second Note."

TWO

Semiotic Things

> No chain is homogeneous; all of them resemble, rather, a succession of
> characters from different alphabets in which an ideogram, a pictogram,
> a tiny image of an elephant passing by, or a rising sun may suddenly
> make its appearance.
>
> —DELEUZE AND GUATTARI[1]

Bruno Latour has dedicated one of his most insightful texts to the gesture
of scientific showing. In "Circulating Reference," he takes the example of
pedology to investigate how a discipline produces and secures the connection between scientific representations and the realities that correspond to
them. How do pedological depictions of things relate to the things themselves? How, for example, are the connections between a map and an actually existing place or between a soil sample and the actual terrain it is taken
from established and secured? "Is the referent what I point to with my finger outside of discourse, or is it what I bring back inside discourse?"[2]

In his remarkable field study, Latour answers symmetrically: neither one
nor the other. Reference, instead, is brought about by a "concatenation of
elements" situated on the boundary between discourse and non-discourse,
elements that are simultaneously matter *and* form, thing *and* sign. In concrete terms: pedology has to establish connections with its object that are

as palpable as they are symbolical. To be able to give a scientific account of the soil at all, pedology has to distance itself from it, rise above it. But at the same time it has to ensure that it always returns to this object. It must not be out of touch, as it were, with that soil.

An investigation of the Helmholtz curves suggests a similar notion. There, too, the act of showing rests on an *agencement*, a concatenation or chain that leads from one form of matter to another, from one semiotic *thing* to another. But it also leads through different semiotic *spaces* and thereby connects the organic with the nonorganic, the ideal with the real, the singular with the universal. On the one hand, it is necessary for emphasizing the scientific nature of one's activity that one achieve this concatenation in a way that can be reproduced or at least be understood. On the other, this concatenation also guarantees the connection with the object under investigation. In Helmholtz's time experiments, the construction of this connection is part of his argument.

The prominent position the physiologist assigns in the "Second Note" to the act of *démontrer*, showing, is first of all salient in the fact that he added (or had someone add) examples of curves to the manuscript text. These additions endow the showing with an outright physical quality. The physical aspect is also apparent in the additional paragraph in the manuscript of the "Second Note" left out of its publication in the *Comptes rendus*. Separated from the rest of the text by a horizontal line, this paragraph talks about "placing" examples of curves "before the eyes of the Academy" (*mettre sous les yeux de l'Académie*)—as if this institution really possessed a body or at least sense organs.[3]

Nonetheless, "placing" sounds much easier than it was in this case. It is easy to see how the curves could be scratched into the layer of soot on a rotating cylinder in Königsberg. But how were the curves to get to Paris to be placed before the Academy? By making a drawing of them? But that would have meant mediating the showing by interpolating a human actor; it would not have been a direct "placing before the eyes." Or should Helmholtz simply send the cylinder to Paris? But the cylinder was made of glass, and it was unlikely to survive the journey unscathed. The same would have been true for a daguerreotype, provided one could have been produced that was light enough and magnified the curves sufficiently. So how were the

Helmholtz curves to get to Paris? Were there "immutable mobiles" that could have safely conveyed them?[4]

The solution to the problem lay halfway between manual drawing and photograph, between paper and glass. Helmholtz used a transparent, adhesive film to detach the layer of soot, with the traces scratched into it, from the cylinder. The film consisted of "fish glue," a transparent substance obtained from the lining of swim bladders, in particular from sturgeons. Since the eighteenth century, cleaned and dried swim bladder linings (usually obtained from Russian beluga sturgeons) had served in the clarification of wine and beer, in the production of putty for glass and porcelain, and in the fabrication of artificial pearls. Processed into thin layers, however, fish glue was also used for tracing images or for obtaining "sharp prints" of coins. Lithographers and engravers used this transparent support medium, also called *Leimfolie* (lit., glue film) or *Glaspapier* (lit., glass paper), for reproductions. They placed sheets of fish glue on the drawing to be reproduced and traced its outlines with a dull needle. They then applied pulverized graphite or red chalk to this retraced image, placed it image-side down onto a stone or a copper plate and, applying pressure, produced an imprint.[5]

It is possible that Helmholtz's time as teacher at the Berlin Academy of Art had familiarized him with this procedure. In any event, he used fish glue in a similar manner to transfer his curves into a medium that enabled them to be transported. He used it to make "imprints" (*Abdrücke*)[6] that he mounted on thick paper or cardboard. In the already mentioned additional paragraph in the "Second Note," this process is described as follows:

> I succeeded in conserving the soot together with the curves the stylus traced in it by rolling the outer face of the cylinder across the moistened surface of several well-bonded and very transparent sheets of fish glue [*colle de poisson*] of the kind used by engravers for tracing. I then glued the surface of the sheet that had taken the soot off the cylinder onto paper.[7]

So this is how the "filmstrips," which have been preserved to this day in the archives of the Académie, were produced.[8]

What seems remarkable about the way these strips were made is that it preserves, at the core of a new method of showing in modern physiology, one of the most archaic procedures for producing images: the imprint, *l'empreinte*.[9] Latour has emphasized the discontinuity within the referential chains of science, the leaps from one element to another. Helmholtz's curve imprint embodies the gap between experimental setup and written message, between technique and text. Yet at the same time, it establishes a solid connection between the two. The imprint is a bridge rather than an abyss.

The transparency of the supporting medium suggests that we look in two directions: both vertically, back to the material preconditions of the production of images or signs in the Königsberg laboratory, and horizontally, at the pictorial or semiotic *milieu* of the Paris Academy into which the results of this production inserted themselves and in which they spread. As we have said, the curves fixed on the filmstrips—on the one hand—represent the palpable connection with Helmholtz's experimental setup. A remarkably organic surface, these strips lead from the paper, the manuscripts preserved in the Académie archives, back to the tables, supporting frames, and instruments that Helmholtz had combined and confronted with one another at Königsberg University to be able to take his time measurements. But the filmstrips do not only take us to the technical, in the widest sense, elements of the experimental setup. They also take us to its organic components.

At one end of Helmholtz's drawing machine there are fish whose swim bladders have been cleaned, dried, pulverized, and, finally, moistened again to bind traces of soot and transfer them from a cylindrical to a flat format. At the other end, we encounter frogs whose calf muscles are suspended from a frame in such a way that their contractions can scratch traces into the soot applied to a rotating cylinder. It is only thanks to this astonishing encounter between fish and frog and a sophisticated chain of semiotic things placed between these organisms that it was possible in the first place for these curves to be admitted, *as curves*, to the convolute that is to this day associated at the Académie with the "Second Note."

As we have seen, Helmholtz was concerned with establishing a connection as convincing as possible between his laboratory and the assembly room of the Paris Academy. Yet the apparently simple "placing [the curves] before the eyes" of the Academy has a correlate in a complex chain, as material as it is semiotic, in which the organic and the mechanical, the imprinted and the handwritten, the indexical and the symbolical, connect with one another and refer to each other to emphasize, we might say, the facticity of a physiological fact.

The references of the heterogeneous assemblage Helmholtz mobilized to this end go even further. The connection between the "Second Note" and the experimental setup established by the "glass paper" does not finish in Königsberg. From there, the rotating cylinder, which consisted of a champagne glass, and the fish glue, which was obtained from that purveyor of caviar, the sturgeon, put us on a semiotic track that leads all the way to Russia and the Caspian Sea . . .

There was no way of telling all this simply by looking at the curves. Quite the contrary. Once they had been detached from the experimental setup and sent to Paris sometime during the summer of 1851, they gave a decidedly compact impression. The view they offered, in any event, was by no means sweeping and grand but rather small and exquisite. That much we can gather from a historically comparative look at curve sizes.

A 1940 photograph shows Walter B. Cannon in his physiology lab at Harvard, sitting in front of a kymograph recording of several meters' length, on which, even at three meters' distance, curves are clearly visible (Figure 8). The curves are about the same size as the glasses through which Cannon studies them attentively. It appears that the system of physiological curve recording has reached maturity here. Although Cannon uses a kind of pen to mark the curves (something that Helmholtz had already done as well, as we'll see), there can be no doubt that there is nothing artisanal about this. It's large-scale industry; it can draw on a maximum of specifically prepared machine performance.

Fifty years earlier, in the heyday of Marey and the graphical method, there were no curves of comparable dimension yet, and kymographs with fanfold paper did not yet exist either. But already at that time, physiological recordings were usually made on drums about as high as today's letter-sized sheet of paper and with a diameter roughly equivalent to the width of a postcard. And Carl Ludwig's kymograph, the one he presented in his pathbreaking 1847 publication, had comparable dimensions: a height of about 25 centimeters (about 10 inches) and a diameter of 15 cm (about 6 inches). Helmholtz, by contrast, was working with a significantly smaller

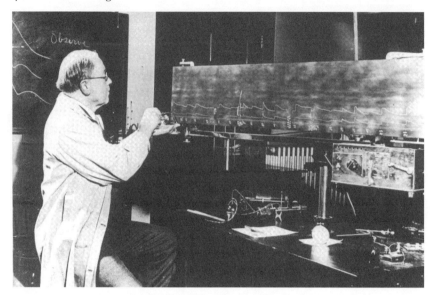

FIGURE 8. Walter B. Cannon in front of curve recordings in Harvard University's physiology laboratory (1940). The kymograph Cannon uses here is a machine with two drums connected by a paper loop of several yards' length. The size of these drums differs little from that of recording cylinders in the 1880s but significantly from that of Helmholtz's machine, which was much smaller. Reproduced from John Parascandola, Toby A. Appel, and Daniel L. Gilbert, eds., *A Century of American Physiology* (Bethesda: National Library of Medicine, 1987), cover illustration. Courtesy US National Library of Medicine.

cylinder that, furthermore, was not made of brass (as was usually the case) but of glass. The diameter of this tube was only 2.7 cm (just above 1 inch)— confirming its origin in a champagne flute.[10]

The dimensions of the curve images were of corresponding size. The dark gray surface on which the curves lucidly stand out was 8.4 cm wide and 2.8 cm high (3.3 and 1.1 inches, respectively). Ludwig had called the recordings of a horse's circulation and breathing movements he had made on December 12, 1846, a "first stammering of heart and breast."[11] In this description, the analogy with language seems to take into account the use of ink and paper that was characteristic of Ludwig's recording device. It also indicates the rather unhurried pace of his kymograph. For one turn around its axis, Ludwig's device needed about a minute—65.5 seconds, to be precise. Not so for Helmholtz's machine, whose glass cylinder made six full rotations in one second—a velocity that definitely exceeded the speed of any kind of stammering. An additional difference from Ludwig was that Helmholtz's device recorded by scratching into a layer of soot and in a direction opposite to that in which texts were usually read. The beginning of the recording is situated on the right, its end on the left—as if the eye was only to look back at the tracks of a time passing at lightning speed.

Their exquisite compactness was another reason the sheets of fish glue were an intermediary step and not a final stop in the transition from the materially semiotic to the semiotically material characteristic of Helmholtz's "new method." Before the curves produced in Königsberg could be placed before the Paris Academy, they now—on the other hand—had to be transferred into the milieu of more familiar drawings, signs, and writings. The first step in this process was to affix a key or legend, that is, literally, "something to be read." Although of limited dimension, the cardboard pieces on which the glue films were mounted offered a surface appropriate for the purpose.

First, for numbering: "No. I" and "No. II" below the filmstrips, and "1.," "2.," and "3." in the left and right margins for the curves, arranged on both filmstrips on three levels (see Figures 1 and 2). Then, also below the images, two arrows pointing from right to left, drawn to indicate the direction of the process recorded. And then, perhaps most important: capturing in words what one saw or was supposed to see. For "No. I" we read, "Courbes autographes d'un muscle [Autographic curves of a muscle]," for "No. II," "Courbes doubles autographes d'un muscle [Autographic double curves of a muscle]." This is the first time in the entire convolute of manuscripts pertaining to the "Second Note" that the term is not "drawing" or "trace" but "–graphic"—that is, writing.

"Autographic" is to be understood in the sense of "written in one's own hand" (a remarkable expression given that we are talking about a frog muscle). But it is also to be read as "self-drawing" or "self-registering" (as in, for example, "phonautograph"). The expression can also be seen as a reference to a printing procedure widely used in the early nineteenth century: autography.

In the present case, however, it was based not on paper and a special ink but on the use of fish glue. *Courbe autographe* would then mean something like "curve imprint."[12]

The last addition affixed on the cardboard piece was a signature. In both cases, the name "Helmholtz" is written in the lower right-hand corner of the card, evidently to bundle the authorship of the individual components in a proper name and to prepare them for being tied into a scientific text that has also had an author ascribed to it.

Transpositions and transformations

Mounting and labeling the curves was the beginning of further transpositions and transformations. A one-page text enclosed with the curves, for example, was to serve as an "Explication des épreuves," an explanation of the curve samples. The text shows the curves once more, albeit in schematized form. This schematization, however, had already begun on the glass cylinder. The Königsberg drawing machine allowed not only for curviform recordings but also for adding marks in the form of horizontal or vertical straight lines. When the stylus was placed on the soot-covered rotating cylinder but no contraction was triggered, it produced a sort of baseline. If, by contrast, a contraction was triggered but the cylinder was not set in motion, the result was a mark perpendicular to this baseline. The text of the "Second Note" describes the horizontal straight lines as "abscissae" and thereby turns the vertical lines into ordinates.[13] And indeed, the vertical and horizontal markings make it seem as if the recorded curves had been inserted into small cutouts from a coordinate system.

Helmholtz, to be sure, at no point tried to provide an algebraic formula for his "autographical curves." Nonetheless, mathematical associations are not coincidental here. They are an expression of a scientific ideal according to which the only way of doing justice to the phenomena of life in a scientific manner is to subject them to a "physico-mathematical treatment." The decisive characteristic of this procedure, as du Bois-Reymond had emphasized time and again since the late 1840s, consisted in "representing to oneself the causal connections of natural phenomena in terms of the mathematical image of dependence."[14]

Now, on the level of representation, the decisive element of the image of dependence is not a numerical value, as one might think. Instead, it is a depiction based on "calculated and observed numerical values" and then presented as the "image of a curve."[15] One of the explicit models here was the work of Adolphe Quételet, who in his 1836 *Sur l'homme* had represented the dependence "of literary talent on age," for example, in the form of a curve.[16] When du Bois-Reymond, in 1848, speaks of "curves," he does not seem to have an instrumental practice of self-recording in mind. He rather seems to be talking about a theoretical instrument, an image-tool, to illustrate the relation of an "observed effect" to "several variable circumstances," that is, the relationship between dependent and independent variables.[17]

The schematic manual drawings that stand at the beginning of Helmholtz's "Explication des épreuves" are a concrete approximation of this mathematical understanding of curves. First of all, these drawings do not present us with three curves arranged one above the next but limit themselves to reproducing a single idealized form. At selected points, this ideal curve is marked with Greek letters, check marks, and lines, as if one were trying to find out "the sense of its inflection toward the abscissa."[18] Next, we find elaborations on the short legends of the cardboard pieces. In reference to "No. I," for example, we now read: "Three curves described by a single contraction of the muscle and at a lower rotational speed"—as if one had already looked at "No. II" or were otherwise capable of understanding what the comparison refers to.

Next comes an explanatory passage that includes the Greek letters:

> The lower horizontal line is the axis of abscissae, the vertical α marks the point that corresponds to the moment at which the nerve is stimulated or to the origin of the abscissae, $\alpha\beta$ the time lost, $\beta\gamma$ the ascending branch of the curve, $\gamma\delta$ the descending branch. The waves between δ and ε correspond to the vertical waves of the stylus, which are the product of the inertia of the metal pieces set in motion and the elasticity of the muscle.[19]

This separates the curve into different sections. These sections correspond to conceptually delimited physiological processes. Beyond that, the explanation introduces a distinction between fact and artifact: on the one hand, the authentic drawing in the first part of the curve (on the right); on the

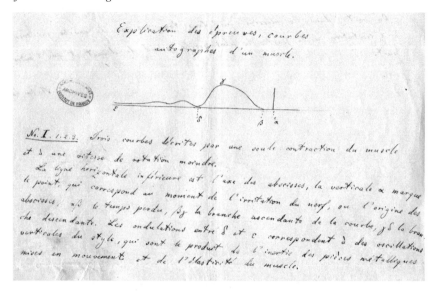

FIGURE 9. Schematic presentation of Helmholtz's "Curves No. I" (fig. 1) in the accompanying note, "Explication des épreuves" (1851). In the note's drawings, Greek letters serve to distinguish individual phases of muscle contraction. Here, the distance between (α) and (β) corresponds to the *temps perdu*. Reprinted with permission from Académie des sciences—Institut de France, Paris.

other, the waves that are due to the reverberation of the recording device and do not really belong to the drawing of the muscle contraction (Figure 9).

The procedure is the same for "No. II" (Figure 10). First, we are given an idealized drawing, then an elaboration of the legend. This time, the explanation does not contend itself with providing a comparative description of the performance of the machine that drew the curves. Now it also offers quantifications:

> Three double curves, which were traced at a greater rotational speed of six rotations per second and were traced by two muscle contractions, I achieved by successively stimulating two parts of the nerve whose distance along the nerve was 55 mm.[20]

Finally the explanatory text that includes, besides the letters, the check marks as well:

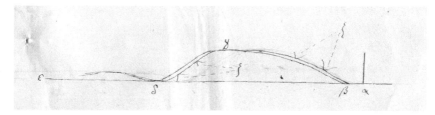

FIGURE 10. Schematic presentation of Helmholtz's "Curves No. II" (fig. 2) in the accompanying note, "Explication des épreuves" (1851). Reprinted with permission from Académie des sciences—Institut de France, Paris.

> The entire length of the drawing corresponds to the circumference of the cylinder and thus amounts to ¹/₆ of a second. The marking ςς [which, via lines, points to the check marks] serves to distinguish the curve executed first from the other. I have scratched them [the check marks] with a stylus on the surface of the cylinder before I traced the second curve. In those cases in which I first stimulated the part of the nerve closer to the muscle, they are at the front of the curve (Figures 1 and 3 [in Helmholtz's curve recordings]); where it was the more distant part, in turn, they are at the end. One sees in the drawings that in all these cases the curve that precedes is the one that corresponds to the stimulus in the more proximate part and that the other [curve] follows it.[21]

And indeed, enlargements make it possible to discern tiny marks that Helmholtz applied with his own hands to the soot layer and that are therefore also preserved on the filmstrips (Figure 11).

These markings recall a central problem of the "new method," the assumed identity of the repetition. Only on the condition that the two successive curve recordings differed merely in the changed location of the stimulation of the nerve—on the condition, that is, that everything else remained the same—could one presume that the interstice between the curves was due to the time needed by the stimulus to move from one point in the nerve to the other.

Simple markings on the curves, to be sure, could not really show this. But they were able to show that, independent of the order of the successive experiments, there appeared in each case an interstice; that is, there was an interim. Thus, independent of whether it was a point on the nerve located

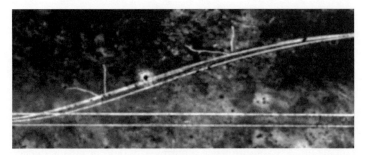

FIGURE 11. Enlarged detail from Helmholtz's "Curves No. II" (fig. 2) (1851). This is the left section of the middle double curve (Helmholtz's "fig. 2"). Reprinted with permission from Académie des sciences—Institut de France, Paris.

at great distance from the muscle that was stimulated first or whether, inversely, it was a point at close proximity to the muscle that was stimulated first, in both cases the two curves shifted in relation to one another. The identity of repetition becomes effectively palpable here as a kind of empty center around which the recording practices of the "new method" were organized.

Neither the curves nor the "Explication des épreuves" were ever published in the *Comptes rendus*, a journal that did not provide any space for illustrations until the 1870s. And yet there can be no doubt that showing a physiological fact that had previously been ascertained and communicated was a central motivation not just for the "Second Note" but for Helmholtz's deployment of the curve method as a whole. In fact, the problem of showing stands at the beginning of the project of constructing (or of having someone construct) a drawing machine. Thus, for example, Helmholtz writes to du Bois-Reymond on September 17, 1850, one year before submitting the "Second Note" to the Académie, that he is

> now having an apparatus with a rotating cylinder built for drawing lines
> with which I hope, among a number of other things, to place before
> everyone's eyes, in 5 minutes, the fact of the duration of propagation in
> nerves in an experiment. I plan on traveling with this [apparatus] to German
> [i.e., German-speaking] universities next summer and to give
> performances.[22]

The "Second Note" seems to echo this passage not just once but twice: first in its emphasis on the temporal aspects of experimentation and demonstration, second in the "placing before the eyes," the *mettre sous les yeux*.

As for the announcement of traveling from university to university, that is exactly what Helmholtz did in August and September 1851. While, in Paris, the "Second Note" was received, read, and processed, Helmholtz visited physiological teaching and research centers in Heidelberg, Zurich, Vienna, and other cities to do one thing above all: to show. To Olga, who was in Berlin during these two months, he writes: "My frog curves, I demonstrate them . . . everywhere."[23] During this time, he also sent the products of his work with curves to Alexander von Humboldt—for no other reason than to "impress" the Berlin savant "with something to look at."[24]

The particularity of the Helmholtz curves, then, consists in their renewed representation of the results of measurements already taken in a different medium. In Peter Galison's terms, one could say that here, a "mimetic" experimental technique (the method of curves) becomes the form in which a "statistical" technique (Pouillet's method) is represented.[25] As we shall see in more detail, however, this relationship between image and logic—which, in this case, also appears as the relation between the analog and the digital— is not a simple, stable one. For, inversely, it had been curves that had also set in motion the taking of measurements according to Pouillet's method. The "new method" described in the "Second Note" was, in this respect, also an old method, or at least one with which Helmholtz had, in a slightly different form, experimented before.

Helmholtz's 1851 return to the method of curves indicates how strongly, in this case, the so-called internal conditions of scientific research are connected with its so-called external conditions. There is no doubt that with his "new method," Helmholtz reacted to a central problem of scientific production and communication: To accomplish a scientific achievement, it is not enough to accomplish it. As historians of science have emphasized time and again, the achievement has to be known and acknowledged as such. Yet obtaining this notoriety and acknowledgment is only partially up to the individual scientist.[26] To a considerable degree, it is up to the scientific community—even if, as in Helmholtz's case, this community is only just emerging.

In what follows, we will look at the obstacles placed along this process of acknowledgment. Through a detailed reconstruction of the scientific, technical, and semiotic aspects of the research process that Helmholtz had been engaged in since his move to Königsberg—a process that may not have culminated but certainly returned to itself in the curves he sent off to Paris in the summer of 1851—we will come to understand what these obstacles were.

THREE

A Research Machine

> The interval is an operation.
>
> —MARCEL DUCHAMP[1]

Let us, therefore, go back to the beginning. This beginning does not correspond to a point but to a "tangle," a "multilinear ensemble" in which *all* lines—not just the lines of art but the lines of technology also—are "subject to changes in direction, bifurcating and forked."[2]

The winter of 1849–50 was Helmholtz's first winter in Königsberg. Together with his wife Olga, he had moved from Potsdam to the city of Kant at the end of August to become Ernst Brücke's successor at the university there. After his studies in medicine under Johannes Müller in Berlin, Helmholtz had spent some time working as an assistant in the Anatomical Museum, which came with teaching assignments at the Academy of Art. The Königsberg professorship in physiology and anatomy was his first major university position. The focus of his 1842 doctoral dissertation had been on anatomy. Evidently prompted by cell theory, new at the time, Helmholtz had used the example of invertebrate animals to show that the axons of peripheral nerves are connected to ganglia.[3]

Subsequently, he moved more and more toward experimental physiology. This field, just then emerging, initially led him to model his research on chemistry, then, later, on physics. His first research project after the dissertation was dedicated to fermentation and putrefaction processes, which he—in opposition to the vitalism of his time—traced back to a purely chemical mechanism. In 1845, he undertook experiments on the "consumption of matter," and in 1848, on the "generation of heat" in muscle activity. In between, he wrote a comprehensive overview of "Heat (in physiology)" and gave his programmatic lecture on the conservation of force (published as *Über die Erhaltung der Kraft*).[4]

In 1849, then, the move: Königsberg instead of Berlin, which seems to have meant time measurements instead of temperature measurements, nerve stimulation instead of muscle action. In fact, both applied. At his new location, Helmholtz was to be concerned with heat *and* time, with muscles *and* nerves, with physiology *and* psychology.[5]

On January 15, 1850, he concluded work on the first research paper of his Königsberg period, entitled, "Preliminary Report on the Propagation Speed of Stimulations in the Nerve." The object of the report, whose printer's copy had been handwritten by Helmholtz's wife, was the time measurements he had taken since October 1849 on the motor nerves of frogs. The opening lines summarize the result of this labor:

> I have found that a measurable time passes while the stimulus exercised by a momentary electrical current on a frog's sciatic plexus propagates to the point where the thigh nerve enters the calf muscle. In larger frogs, whose nerves were 50–60 mm long and which I had kept at a temperature of 2–6°C, while the temperature of the observation room lay between 11 and 15°, this time was between 0.0014 and 0.0020 of a second.[6]

The research machine that stands behind these sentences was based on a telegraphic connection between a frog's lower leg, a galvanometer, and a telescope in the Königsberg University building. The time measurements, made public for the first time, were thus not the result of a graphical procedure but had been obtained by using a basically electromagnetic method.

This method made time visible without fixing it; it did not write it as a curve, did not transform it into a trace, but rather—not unlike a clock—made it readable on a scale, such that it could then be recorded in the form of numerical values.

The complicated nature of this research machine recalls Duchamp's *The Large Glass* or the involved assemblages described by Raymond Roussel in *Locus Solus*, the *demoiselle à rêître en dents* (roughly, "paving-beetle with a mercenary in teeth"), for example.[7] It is temping to draw (and to draw the map of) this machine, as Jean Ferry has done for Roussel's device.[8] For obvious reasons, I'll take a different approach. In what follows, I give a description, enriched with quotations and historical depictions, of the different areas of the research machine that Helmholtz not only made use of when he performed his psychophysiological experiments but also had put together in the first place. This description goes into great detail but also structures these details into areas. Given the difficulties that already plagued contemporary scientists when confronted with Helmholtz's way of describing his methods and procedures, this description cannot lay claim to comprehensive clarity. In this respect, too, it is a challenge for scientific and technical imagination.

Helmholtz based his precise time measurements on electromagnetism. Before moving from Berlin to Königsberg, he had constructed a "frog drawing machine"[9] that was to mechanically record muscle movements in the form of curves. He does not, incidentally, speak of "graphical method,"[10] a name that was not adopted until the 1860s. Helmholtz uses other expressions with geometrical as well as artistic connotations; in particular, he speaks of producing "curves," "lines," and "drawings." Once in Königsberg, he turns his back—at least temporarily—on the techniques and procedures at the basis of such productions. As late as July 1848 he had written from Potsdam to his then-fiancée Olga von Velten that the drawing machine he had constructed produced curves that were "very fine and regular."[11] Over a year later, in October 1849, however, he writes from Königsberg to his friend du Bois-Reymond in Berlin that he has taken up his investigation of muscle movement again, "following an entirely new method."[12]

The "Preliminary Report" names this new method: the "method indicated by Pouillet for measuring small portions of time."[13] The method, described by French physicist Claude Pouillet in 1844, consists in using the deflection of a galvanometer's needle for the purposes of precise measurements of time. A detailed description of the physiological application and adaptation of this procedure, however, had to wait until the "Measurements Concerning the Temporal Course of the Contraction of Animal Muscles and the Propagation Speed of the Stimulation in the Nerves." Helmholtz submitted this ninety-page treatise for publication in the prestigious Berlin journal, the *Archiv für Anatomie, Physiologie und wissenschaftliche Medicin*, in June 1850, a few months after the "Preliminary Report."

According to the "Measurements," the machine Helmholtz used for his precise time measurements was organized into three large areas. First, the frog frame. This frame was used to suspend the dissevered calf muscles of frogs, to attach wires and weights to them, and to make them contract (Figure 12). To this end, the motor nerve was exposed from the upper end of the calf muscle all the way to the sciatic plexus. In the case of the big frogs Helmholtz was working with in Königsberg, this was a distance of 5 to 6 centimeters (about 2 inches). He then hung the muscles by the "lower joint part of the thigh bone" from an adjustable screw, which entered the bell jar covering the tower-like "frame" from above.[14]

Next, the exposed nerve was attached to copper wires at two different points (Figure 13). At these points, it was possible to momentarily stimulate the nerve with electricity from a battery that consisted of Daniell cells and an induction coil. Meanwhile, "wet cardboard disks" ensured that the air in the largely closed space of the bell jar remained moist enough for the exposed nerve to be "effective for 3–4 hours."[15] The prepared nerve-muscle sample had thus been violently separated from the rest of the frog organism in order to live on—or at least be kept in a life-like state—for a little while longer under the bell jar. Not a word is lost about whether, and if yes, for how long, the rest of the organism continued to live.

Instead, work continued on the experimental setup. At the lower end of the muscle sample, Helmholtz hooked in a narrow steel frame that went almost all the way to the bottom of the tower-like wooden frame. This steel frame contained a series of alignment pins, setscrews, and contact surfaces for adjusting the muscle and for establishing connections with a second electrical circuit. At the lower end of the frame, a "dish . . . for weights"

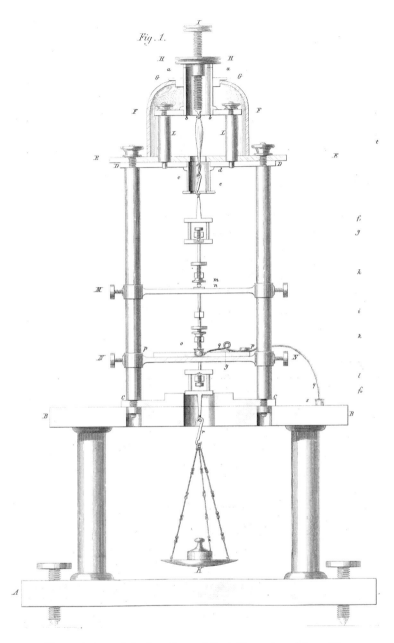

FIGURE 12. General view of the frog frame (1850). Reproduced from Hermann Helmholtz, "Messungen über den zeitlichen Verlauf der Zuckung animalischer Muskeln und die Fortpflanzungsgeschwindigkeit der Reizung in den Nerven," *Archiv für Anatomie, Physiologie und wissenschaftliche Medicin* (1850), 276–364, table.

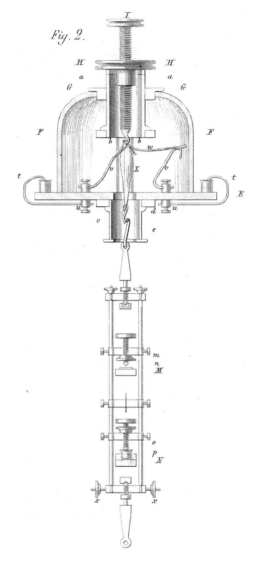

FIGURE 13. Detailed lateral view of the frog frame (1850). Under the bell jar (F), the nerve-muscle sample (L) hooked into the screw (I). The nerve (w) visible at the top of the muscle is connected to copper wires (v). At the bottom of the muscle, the steel frame with several contacts and set screws, including the mercury interface (P) and (N). Reproduced from Hermann Helmholtz, "Messungen über den zeitlichen Verlauf der Zuckung animalischer Muskeln und die Fortpflanzungsgeschwindigkeit der Reizung in den Nerven," *Archiv für Anatomie, Physiologie und wissenschaftliche Medicin* (1850), 276–364, table.

was attached.[16] Thanks to this dish and by means of the setscrews in the steel frame, the muscle could be extended without stretching it.

All in all, this area resembled the experimental setups that had been developed since the mid-1830s by life scientists such as Theodor Schwann and Gustav Valentin for the experimental study of muscle force. More than anything, however, it resembled the time-measuring setups of Carlo Matteucci, who, around 1845, had begun working with hanging frog leg muscles (see chapter 4). Helmholtz, however, does not refer to Matteucci but to the work on "muscle movement" done, also in the mid-1840s, by the Leipzig anatomist and physiologist Eduard Weber.[17] Weber, too, had used hanging frog samples in his experiments; he had used tongue, not leg, muscles. He, too, had attached weights to the suspended muscles and stimulated them with an electrical current. Yet those studies, as Helmholtz explains, were mainly concerned with the actions of the muscle "in the resting and the continually excited state."[18]

The interest of the physiologist freshly arrived in Königsberg went in the opposite direction. The main question for him was how the muscle switched from rest to excitation and, especially, how rapidly this "switch" took place.[19] His basic assumption was that the speed of this switch would yield decisive information about the "mechanical work" of the muscle, about its performance, as we would say today. A frog sample, for example, might be able to lift a 200-gram weight. Yet to give a more precise description of this property, we have to take into account the speed with which this lifting took place: as quickly as a pedestrian carries a load from A to B or as quickly as a locomotive carries the same load over the same distance?[20]

A possible point of departure for Helmholtz's project thus outlined was a thesis of Weber's concerning "animal muscles." Weber used this term to denote "all cutaneous and skeletal muscles; the muscles of the tongue, the palate, of pharynx and larynx, the diaphragm, the muscles of the anus, the penis and those by means of which we can block or allow the discharge of urine."[21] These muscles, Weber held, were characterized, on the one hand, by the fact that they could be activated "arbitrarily." On the other hand, while a stimulus "immediately" causes these muscles' "contraction," this contraction subsides "just as quickly . . . the moment the stimulus ceases."[22] Helmholtz's "Measurements" aimed at investigating this alleged instantaneity more closely and, if possible, to define it more precisely. To come closer to a physics of the muscle, what seemed to be sudden and immediate in the muscle's action was to be grasped as a kind of duration.

That is why Helmholtz used electricity not just for stimulating the muscle but also for measuring time. This was the second area of his research machine: the galvano-chronometer. To take precise time measurements, Helmholtz added a second electric circuit to the first one; the two were punctually connected. The first circuit served only to stimulate the prepared nerve-muscle sample under the bell jar. In the second circuit, in turn, the measurement of "electricity time" was to take place by means of a galvanometer with "1400 coils of wrap spun copper wire."[23] Helmholtz does not say anything about the provenance of this electrometer. It is not clear whether his was a commercially available instrument (Figure 14) or whether he constructed it himself.

The two circuits of the setup were laid out in such a way that a specifically inserted switch, a "rocker," made it possible to close them simultaneously: The sample was stimulated and, at the same time, a current flowed to the galvanometer. The contraction of the muscle then interrupted this flow of electricity again. This, precisely, was the time interval, the "electricity time" to be measured: from the stimulation of the muscle to the onset of its contraction. But how could one take the step from this interruption of an electric current to the measurement of time?

This, precisely, is where Helmholtz drew on Pouillet. Pouillet had started out from the idea that the magnitude of the galvanometer needle's deflection depended not only on the intensity of the electrical current but also on the time during which the current has an effect on the galvanometer.[24] Accordingly, it had to be possible to make inferences from the deflection of a galvanometer needle to "electricity time," especially if extremely short intervals were at issue. The electrometer thus in a way became an electricity clock. To demonstrate this possibility of converting the galvanometer, Pouillet had set up an electrical circuit in which a rotating round disk, on which precisely dimensioned contact surfaces were affixed, could produce electrical impulses that were exactly delimited temporally. He then compiled tables of the needle's deflection at an invariantly strong current

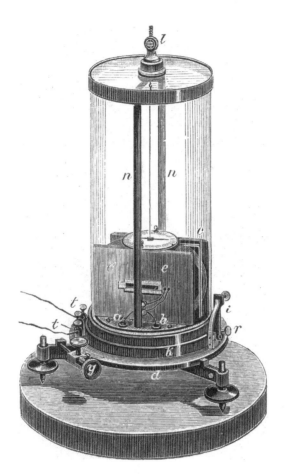

FIGURE 14. Galvanometer (1876). The image shows a galvanometer for electrophysiological precision measurements of the kind du Bois-Reymond developed in the late 1840s. Here, too, the essential components are housed under a bell jar: the coil with its various windings (a, b, c, e) and the magnet bar suspended above it from a silk thread. Reproduced from Elie de Cyon, *Atlas zur Methodik der Physiologischen Experimente und Vivisectionen* (Giessen and Saint Petersburg: Ricker, 1876), table XLIV.

but at intervals of various lengths. It turned out that these deflections were indeed reliable indicators of the very short times during which electricity had flowed.

In his 1844 publication, Pouillet compared his procedure to the ballistic pendulum constructed by Benjamin Robins a good hundred years earlier. He also named ballistics as one of the potential fields of application when he presented his procedure. It is thus no surprise to find illustrations in contemporary manuals on electromagnetism that depict the use of Pouillet's method in the area of ballistics (Figure 15). What distinguished Pouillet's method from Robins's classical ballistic pendulum was this: Whereas the pendulum marked its maximum deflection with a ribbon and thus fixed it, Pouillet's procedure transformed a fast movement (that of the short-interval current) into a significantly slower one (that of the needle) that was in turn amenable to precise observation. The electrical impulse

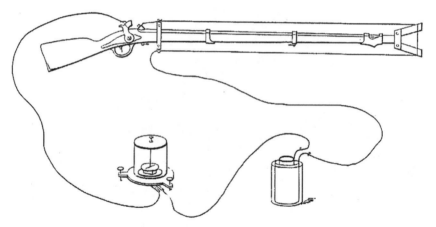

FIGURE 15. Application of the Pouillet method for measuring short intervals of time in ballistics (1852). In this setup, firing the gun closed an electric circuit. As soon as the bullet went through the wire stretched across the muzzle, the circuit was interrupted again. As in the case of the ballistic pendulum, the deflection of the galvanometer needle allowed for deducing the time that had passed. Reproduced from François Moigno, *Traité de télégraphie électrique, comprenant son histoire, sa théorie, ses appareils, sa pratique, son avenir, sa législation*, 2nd ed. (Paris: Franck, 1852), table 13.

was translated into a needle movement. Electricity expressed itself without friction in the language of mechanics.

Seventeenth- and eighteenth-century physiologists had already compared the effect of a muscle contracting to the firing of a cannon ball. Explicitly citing the "mighty and often horrid forces of Gun-powder," the English physician and naturalist Thomas Willis wrote around 1670

> that it must necessarily be supposed some motive Particles are hid in the Muscle, which, as occasion is given, are stirred up according to the Instinct, delivered by the Nerves from the Brain, into motion, as it were with a certain explosion.[25]

In this regard, it makes perfect sense for Helmholtz to take recourse to a method of measuring time developed for ballistics—even if his model of the muscle was a different one, namely that of an elastic ribbon. This he picks up explicitly from Weber: "According to Ed. Weber's studies, a muscle's mechanical properties resemble that of an elastic ribbon of variable elasticity."[26]

Yet elasticity is different from explosivity. This was manifest in practice as well. The dilation that followed the contraction—to continue with the image, the return of the projectile as if attached to a rubber band—was a nuisance. For in order to take time measurements according to the procedure suggested by Pouillet, the circuit had to be closed *permanently* after its short action in the galvanometer circuit, since the needle, for its part, needed time to deflect fully. Closing the circuit as such was not a problem: At the beginning of the contraction, the muscle interrupted a platinum–gold contact in the inserted steel frame that was wired accordingly. What caused problems was the permanence. How was the closing of the circuit to be permanent when the same platinum–gold contact closed again once the contraction diminished and the steel frame was lowered again?

This seems to concern only a detail, a kind of one-way switch. And yet it constituted one of the major problems in adapting Pouillet's method for the purposes of physiological time research. Helmholtz claims in the "Measurements" that he "summoned up more reflection and complicated expedients" to deal with this detail than he did for any other part of his machinic assemblage.[27] The solution he finally came up with consisted in inserting an additional contact into the suspended steel frame, underneath the platinum–gold contact. This new contact consisted of an "amalgamated" copper tip placed a little above a small cup filled with mercury. Lifting the cup a little before the beginning of an experiment such that the copper tip was immersed in the mercury and then carefully and slowly lowering it led to the formation of a kind of cone that conducted the current. The beginning of the muscle's contraction caused by the electrical stimulus disrupted the mercury, which assumed its round surface again and remained below the copper tip. The circuit was interrupted permanently, and the electricity time was identical with the interval between the beginning of the stimulus and the beginning of the contraction.

The connection between the first two areas of Helmholtz's research machine, between the frog frame and the galvano-chronometer, was thus not a punctual connection. It was a cone- or drop-like wet interface formed by nothing other than the surface tension of the mercury (see Figure 13).

Finally, the third area of the research machine: the telescope for reading. The telescope brings out the central feature of the entire installation: namely, that it is geared toward seeing. In fact, Helmholtz had assembled a *machine for seeing*, not a writing or drawing machine. The basis for determining time with this machine was reading off the needle's deflections on the galvanometer. This reading did not simply take place with the naked eye and off the scale on the instrument but with the aid of a telescope, a small mirror, and a scale affixed on the outside. A "small magnet rod" of nine centimeters' length, suspended from "fibres of raw silk," functioned as needle or hand of the galvanometer. Above the middle of this rod and perpendicular to the silk threads, Helmholtz had affixed a "little mirror," and vis-à-vis this small mirror, he had set up a telescope. Above or below the telescope, he then mounted a fixed "horizontal scale," whose divisions pointed forward, toward the mirror.[28] Through the telescope, he could then observe the image of the scale appearing in the mirror and follow its movements.

In the design of this area, an area he describes only in piecemeal fashion and gives no illustration of, Helmholtz took recourse to a known procedure as well. It had been described, developed, and used in physics in the 1820s by Johann Christian Poggendorff and some years later by Carl Friedrich

Gauß and Wilhelm Weber (one of Eduard's brothers), albeit in a different context. In both cases, for Poggendorff as for Gauß and W. Weber, the goal had been to measure the earth's magnetic field in precisely circumscribed places. As Gauß and Weber explained, these determinations were to help establish a knowledge of

> very great practical interest. To the mariner, and the surveyor, it must be of considerable importance to know the frequency and magnitude of the disturbances to which the compass is liable, even were it only to learn what degree of confidence he might place in its indications.[29]

This oriented their research in the Magnetic Association, founded in 1836. The goal of this association was to have terrestrial-magnetic measurements taken in as many places as possible by observers as educated as possible to ensure the greatest possible reliability, then to collect them in Göttingen, process them, and publish them.

Expensive and complicated instruments, like "Gambey's compass," which von Humboldt had used, were to be avoided. Ultimately, the matter of costs had already motivated Poggendorff's 1826 suggestion to work not with a compass but with a large magnet bar and a telescope. Extended substantially by the question of exactitude, this is also the backdrop for the *Results of the Observations Made by the Magnetic Association* that, ten years later, Gauß and Weber would start to publish annually. Therein, they made concrete suggestions for setups, gave directions for taking measurements, and suggested budgets for terrestrial magnetism observatories. In addition to the detailed text, they included illustrations of instruments and expedients to ensure "that any clever artist can work from [them] with certainty," that is, to ensure that the standards they announced could be applied effortlessly in the most remote provinces.[30]

The procedure that guaranteed that measurements would yield precise results, the method of reading off the mirror, was described in great detail. Unlike Poggendorff's mirror, which had been affixed on the magnet bar "in the middle and roughly in parallel with its axis," the Göttingen magnetometers' mirror was "fixed to that end of the magnet bar which is turned towards the telescope."[31] In a room free of all metal, a wooden house, the magnet bar was suspended from silk threads hanging from the ceiling. To protect it from external influences like drafts of air, it was encased in a wooden box that had an opening at the front. The telescope and the scale were set up at five meters' distance from the bar. According to Gauß, this had the great advantage that "the observer always remains at a great distance from the needle" and does not "adversely" affect the needle "through his own heat" or through "iron or even brass" he might have on him (Figure 16).[32]

FIGURE 16. Gauß's observatory for precise measurements of terrestrial magnetism in Göttingen (1837). At the center, the "magnetometer," a long magnet bar suspended from the ceiling on silk threads. A wooden box protects this bar from air movements in its vicinity. At fixed times, the position of the magnet bar is precisely determined with a telescope mounted on a tripod (front right). It aims at a mirror affixed to the front end of the magnet bar such that the mirror reflects the scale affixed at the bottom of the tripod. To the right of the telescope, a pendulum clock for the temporal coordination of observations. Reproduced from Carl Friedrich Gauß and Wilhelm Weber, eds., *Resultate aus den Beobachtungen des magnetischen Vereins im Jahre 1836* (Göttingen: Dieterich, 1837), table 1.

In the discussion of potential disturbances, other aspects of the entire installation are examined in great detail, for example the elasticity of the silk threads from which the magnet bar was suspended. Weber dedicated an entire essay of its own to this problem in the *Annalen der Physik und Chemie*, while Gauß, in a famous remark in the first volume of the *Results*, points out that "the presence of larger insects" or of cobwebs in the wooden case that encloses the magnet bar could be an "impediment" to the bar's free movement and therefore "spoil" the observation.[33]

The two Göttingen scientists, however, did not use the mirror reading method only in their terrestrial-magnetism measurements in the Magnetic Observatory (Gauß) and the Physical Cabinet (Weber). They also used it in transmitting and receiving telegraphic messages, which were sent back and forth between the two observation stations in the Observatory and the Cabinet. The goal of this local telegraphic link was, on the one hand, to compare the clocks set up in both locations "more sharply than by any other means" (that is, to synchronize them); on the other hand, "whole words and short phrases" were transmitted in order to harmonize concrete stages of a process (Figure 17).[34]

In fact, the telegraph Gauß and Weber developed and operated in the early 1830s was a kind of miniature magnetometer that was moved by terrestrial magnetism, to be sure, but above all by targeted electromagnetism. At both ends, this telegraph essentially consisted of a large magnet bar suspended on a thread with a mirror mounted in the middle, not unlike Poggendorff's magnetometer. Housed in an elongated wooden frame, the bar was girded on both ends by multiplicator coils.

The deflections of the bar effected by switching power on or off— Gauß speaks of "twitches" and "twitching movements," as if he were talking about a muscle[35]—were observed via a mirror, telescope, and scale, and, with the help of a code, could be translated (back) into letters. In a way, it was a compromise formation between electrical and optical telegraphy. It seems that Gauß had already gained experience with optical telegraphy in the early 1820s during his work on the Hannover arc measurement.[36]

When Helmholtz adapted the mirror reading method from Gauß and Weber, he transformed a partly optical, partly electrical telegraphic proce-

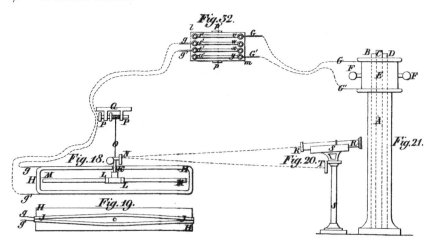

FIGURE 17. Components of a Gauß-Weber telegraph (1838). "Fig. 18" depicts the receiving device. It is magnet bar (M) suspended via silk thread (O) from a set screw (Q), which, as in the magnetometer, is affixed to the ceiling. Induction coils (H) make it possible to move the magnet bar (M) with the transmitter ("Fig. 21"), the commutator, set up at a distance. "Fig. 20" shows the reading telescope mounted on a tripod. It is aimed at a mirror (N) affixed to the middle of the magnet bar. The mirror allows for observing the deflections of the bar on the scale (T) affixed to the tripod, which are translated into messages according to a set code. Reproduced from Julius Hülsse, "Anwendung des Elektromagnetismus auf Telegraphie," *Polytechnisches Zentralblatt* 4/1 (1838): 481–96, table 5.

dure performed by human actors into a telegraphy between human and animal—even if in Helmholtz's setup, only one side sent and received, encoded and decoded: the experimenter. However, this adaptation allowed Helmholtz in his own work to draw on all the theoretical discussions and practical suggestions that Gauß and Weber had published with respect to the magnetometer—from the question of disruptive currents of air via the problem of the silk threads' torque to the determination of the magnet bar's period. Implicitly he thus followed all the indications that first Gauß and later Weber had provided concerning the fact that every magnetometer can also be used as an instrument with which to measure electrical currents "with the utmost sharpness."[37]

By transferring the remarks, descriptions, and explanations concerning the magnetometer onto the significantly smaller galvanometer that, following Pouillet's method, he used as a chronometer, Helmholtz transferred a technical procedure from the domain of geophysics to the newly emerging field of chronophysiology. In so doing, however, he also imported an entire culture of precision into the research domain of the life sciences.[38]

The result was truly remarkable. The research machine put at Helmholtz's disposal a procedure for measuring time that was almost entirely mediated by electricity and light. Neither the telegraphic twitching of the frog muscle nor the deflection of the galvanometer's needle left enduring traces. Instead, a coupling allowed both processes to interlock such that temporal signs temporarily became visible in the form of an image. This image, beheld through the telescope, was that of a thread hanging in front of a scale observed via a moveable mirror. Whereas Helmholtz would later, in a different context, speak of the newer procedures of measuring short intervals of time as a "microscopy of time,"[39] his Königsberg research machine produced, in one respect, the exact opposite: a telescopy of time. Yet in this reversal, the research machine confirmed the fact that the unusually precise measurements of time he was thus able to take were based on a scopic, not a graphical, regime of experimental practice.[40]

More then fifty years later, the relationship of seeing and writing was to be articulated in similar terms by Proust. Of course his novel was a written machine (in just the same way, Helmholtz had to rely not only on seeing but also on handwritten notation to securely process his results). Yet for the reader of *In Search of Lost Time*, the novel was to function as a "telescope, which is pointed at time." In the words of the narrator, Marcel, the emerging novelistic work was to be a tool for seeing "of the sort the optician at Combray used to offer his customers."[41] Using this optical tool aimed at attaining new perceptions, sensations, and thoughts not just in relation to one's own time but also with respect to various (temporal) environments. In that sense, Proust's novel, too, is to be seen as a machine for seeing, even if its precision is of a kind other than numerical.

Helmholtz's research machine, by contrast, was oriented toward measurements, toward numbers. Time measurements were taken as follows. First a point on the nerve farther away from the muscle was stimulated, then one closer to it. In both cases, Helmholtz determined the time needed for the muscle contractions resulting from the stimulations as a whole. The differences between the results obtained, combined with taking into account the length of the nerve in question, then allowed for determining the speed at which the stimulus propagated in the nerve. This was the first concrete manifestation of the variation and subtraction method that was to be so productive later for many studies in the field of psychophysiological research on time.[42]

Unlike Pouillet, Helmholtz in employing this method did not rely on empirical tables to translate needle deflections into units of time. Instead, he set about calculating the time values from the deflections observed. Picking up on the procedure developed by Gauß and Weber, he based his calculations on the general properties of the galvanometer he was using. The decisive variable was the period of the magnet bar. In the 1836 *Results*, Weber had described how to determine this value in the case of small (60 mm) magnet bars. In the 1837 *Results*, Gauß followed with detailed instructions for determining a magnet needle's period.[43] Helmholtz referred to these texts, explicitly citing Weber's 1836 contribution.[44]

On their basis, he first determined the period of the magnet bar he was using. It was 24.607 seconds. The second basic factor was the deflection of the magnet bar that, in this setup, was caused by a steady current. It had to be determined with a telescope prior to every series of tests. Since Helmholtz also took into account "that the magnet [could] never permanently be put in a state of absolute rest"[45] but, due to currents of air and other external influences, was constantly moving, if ever so slightly, he also had to determine, for each observation, how great the deflection was before and after the time-measuring current had had its effect in order to calculate a kind of average, the "meridian." These were the third and fourth factors needed for calculating the time values he sought to obtain. The fifth and final factor was the deflection that took place when the time-measuring circuit was closed, when the contraction was triggered.

There were thus five factors to be considered if precise times were to be calculated from the needle's deflections. Ingenious as this way of proceeding was, it came up against an important limitation. The result was determined not only by the measured contraction times but also by the measured length of the nerve. Yet, due to the "great stretchability" of the nerves, this length was "very uncertain."[46] This uncertainty was one reason the spread in the results concerning the propagation speed of the nervous stimulus Helmholtz reported in January 1850 was not negligible: between 25 and 43 meters per second. The investigations during the months to follow were aimed at more precisely delimiting this result. In the ninety-page treatise of June 1850, the "Measurements," he finally presented an average value of 26.4 meters per second.[47]

Frictions

Against this backdrop—his complex research machine functioning productively and his culture of precision schooled by Gauß and Weber— Helmholtz names concrete reasons for abandoning the curve method. In the "Measurements," he reports that shortly after his arrival in Königsberg, he still pursued the enterprise of investigating the movements of muscles by way of their self-recordings. The model he explicitly claims for this enterprise was the kymographic registration of fluctuations in blood pressure and respiration movements performed by Ludwig. Yet a closer look (see Figures 7 and 8) shows that this was a rather abstract model. Given the speed of the muscle contractions and the orientation toward precise time measurements, the use of a rotating cylinder for a recording surface presented technical challenges that differed from those of Ludwig's drawing of pulse and respiration curves at a pace that was, by comparison, leisurely.

For Helmholtz's studies, the cylinder had above all to be lighter, and therefore also smaller, to be able to turn at an appropriate speed. In the "Measurements," Helmholtz, writing in retrospect, leaves no doubt that the structure of the "simple apparatus" he had built for the purpose was provisional. The purpose of the apparatus had been to furnish practical experience in order to then, on the basis of what had been learned, "be able to construct a definitive" apparatus.[48] Similar in this respect to his Potsdam frog drawing machine, the device in question had made it possible to register the movements of a muscle sample, prepared and with weights attached, by means of a "little steel rod." This registration took place "either on a horizontally progressing, slightly soot-blackened glass plate or on a rotating cylinder surface."[49] It resulted in traces of muscle contractions that simultaneously constituted markers of time: "In this way, the twitching muscle drew curves whose horizontal abscissae were proportional to time [and] whose vertical ordinates were equal to the lifting of the weight."[50]

Helmholtz emphasizes that this procedure had led to "very delicate and very precise curves."[51] In fact they were hardly visible to the naked eye. Their presentation in a scholarly article had to be, accordingly, all the more elaborate. First, Helmholtz chose one "of the lines drawn on the small soot-blackened glass plate" as one that was particularly fit for such depiction. Then he used a microscope to assist him in copying the line in question by hand. At this point, we may literally speak of a microscopy of time. During this copying, Helmholtz also ensured that the curve's "vertical peaks were enlarged 6 ²/₃ times."[52] Thus reconstructed as to its "shape [*Gestalt*]," the drawn curve was given to an engraver in order to finally be printed in the plate accompanying the "Measurements."

Given the painstaking labor involved in depicting this curve, it seems understandable that, in discussing it, Helmholtz kept to its "general form."[53] All this form allowed one to do, he starts by saying rather soberly, was obtain an "overview of the progress of the twitching."[54] Then the turnabout. This overview, he continues, already sufficed to establish—in opposition to Eduard Weber—the "fact, unknown so far" that a muscle's "energy" does not completely develop in the moment of its punctual stimulus "but for the greater part only after this [stimulus] has already ceased."[55] This was the decisive insight already conveyed by the work done with the simple curve machine: The muscle did not, as Weber had said, "instantly" enter "into contraction" thanks to the stimulus. Instead, time passed between stimulation and contraction—not a lot, but enough to be clearly recognizable. The immediacy assumed by Weber turned out to be in fact an interval, a time span, a space of time as delimited as it was empty—an "interim," a *temps perdu*.

But that was not all that could be seen in the curve retraced with the help of the microscope. The reconstructed line as a whole showed that the muscle's "energy" gradually increased, then quickly reached a maximum and finally decreased again slowly. The ascending part also showed, however, that the muscle did not continually lift the weight up; it was not subject to a linear acceleration but instead worked with a series of successive pushes and thrusts. Accordingly, the curve in the first section shows a sequence of concave and convex points. Put differently, the muscle did not simply pull up a weight but quivered and shivered it up. For Helmholtz, this subsequently became the basis for subjecting muscular action to a segmented analysis, for taking measurements at different points in time, in different phases. More precisely, he reconstructed the specific dynamics of these phases by incremental increases in the weight (from 50 to 100 grams, from 100 to 150 grams, and so on) in order to then put together the process as a whole as if from individual images. In this sense, the deployment of the Pouillet method can be described as cinematographic, even though it did not record any movements but only provided numerical snapshots.

Although the curve printed in the "Measurements" could be divided into intervals of 0.10 seconds each, Helmholtz had quickly realized that no more precise time measurements could be taken this way. The main reason, he claimed, was the "influence of friction."[56] What later proponents of the graphical method were to see as its greatest advantage—namely that "the motion to be depicted *itself* marks all of its changes, even the fastest and most temporary ones"[57]—now presented itself as an obstacle to precise time measurement. After several experiments in this direction, Helmholtz, in any case, was convinced that what characterized the curve method, the contact between the object investigated and the recording device—in Peircean terms, its indexical character—could have but one result, namely that "the movement presently taking place [is] slowed down."[58] The "unavoidable friction of the apparatus' individual parts" thus rendered the method useless for precision time measurements. One consequence of this was that physiological self-recording was seen as a "preliminary method." Its main value lay in framing and observing the discontinuous progress of the muscle's twitching in its general form and in stimulating further study.

This assessment is one of the reasons Helmholtz's research machine is not the origin of psychophysiological time research but its ramified beginning. In fact, the figure of repetition is deeply inscribed in this beginning. Helmholtz's criticism of the curve method leads to an almost complete retooling of his research machine in the direction of electromagnetism. Little more than the frog muscle remains intact. Around this organic element, almost all components of the machine are replaced or supplemented with new ones. The transition this entailed from chronography to chronoscopy, however, was not an external matter, not a mere question of method, no simple decision in favor of a different technique. What this transition points to, instead, is a switchover at the core of the entire research project Helmholtz had pursued.

This becomes very clear at the beginning of the "Measurements," where Helmholtz writes that, having understood the shortcomings of the curve method, he had "left the path followed up until then" in order to "clear a different" path.[59] This other path, this newly cleared space for experiments, led him to Pouillet's electromagnetic procedures for measuring time but also to a new research subject. Helmholtz moved from the question of muscle movement to the problem of nerve stimulation.

There were two motivations for this drifting. On the one hand, there was the insight, derived from physics, that what allowed experimental access to the "mechanical work" performed was not only the curves drawn by a muscle, however "fine and regular" they might be. In the same way, and perhaps even more importantly, time made this access possible, since it constituted an essential parameter of the muscle's action. On the other hand, what "cleared" the new "path"[60] in the winter of 1849–50, and would finally to lead to the measurement of the propagation speed of the nerve stimulus, was the precision required and, thanks to Pouillet, achievable for registering this parameter. Not until it became fathomable to measure time in the range of 0.0001 seconds did the question arise of where exactly the muscle was to be stimulated in the experiment: in the middle, where it was thickest, somewhere near the top, or exactly where the motor nerve entered? Perhaps at an even more distant point on the nerve?

Ultimately, it is the tentative alteration of this factor that led Helmholtz to a precise determination, by way of variations and subtractions, of the propagation speed in the nerve. The new precision, the higher resolution of the time measurements, therefore not only raised a question about an aspect of muscle movement that, eventually, was of secondary importance from the point of view of the curve method. It also provided a new answer to this question—an answer that, at the same time, changed the question: At issue now were no longer muscle movements but nerve stimulations.

The refrain of the interim

For the three areas of the research machine—frog frame, galvano-chronometer, reading telescope—to function together, its components needed first of all to be properly distributed or, in Helmholtz's phrase, "fit together":[61] setting up the frame, hanging the steel frame, weighing down the sample, wiring the nerve and galvanometer with the galvanic elements, adjusting the muscle height, positioning the telescope, mirror, and scale, and so on.

Beyond this, however, was also the question of temporal coordination. The individual areas of the machine referred back to differently dated material models that had to be combined—and thus changed and readjusted to one another—in the here and now of the Königsberg laboratory: E. Weber's arrangement using a hanging muscle (1846) and Pouillet's time measurements using a galvanometer (1844), as well as the mirror reading according to Gauß and W. Weber (1837) and Poggendorff (1826).

Yet in addition, as Helmholtz explained, using the machine that had been tinkered together this way required taking into account a multitude of aspects at each and every moment: "turning over highly convoluted wire circuits with second-order side-currents, adjusting the muscle, adding the weights, reading off the scale partitions, closing and opening the chain at the right time."[62] To du Bois-Reymond he admits that having to "pay attention to so many things simultaneously" makes him "completely confused"[63]—especially when a "small microscope" was added, which served to determine precisely how high the muscle lifted the weights.[64]

Because of the confusing simultaneity of the setup, but especially because of the small microscope added for taking readings, it seemed advisable to include an assistant in the work. Otherwise, the experimenter would by himself have to shuttle between the muscle sample in the frog frame and the telescope pointed at the galvanometer and, in addition, go through the multiple movements needed for the electrical stimulus. Helmholtz describes the effect of such solitary work to du Bois-Reymond: "There were so many separate movements to be performed in the complicated galvanic wires on the suspension apparatus that, as long as I did everything myself, I bungled [*verprudelte*] every 3rd or 4th measurement."[65]

In the same breath, Helmholtz explains that the aide he finally employed was none other than his wife Olga. The German expression Helmholtz employs to describe his problems—*verprudeln*—might even be a hint to that effect. It means something like "marring a needlework as one is working on it" and would be used in talking about a seam or about knitting.

In his correspondence with du Bois-Reymond, Helmholtz is very explicit. As the work is still going on, he proudly reports to Berlin that in his experiments, Olga functioned as "the most patient and arduous assistant and recorder," who noted down the "parts of the scale" he observed in the telescope.[66] At a later point, he writes she had taken over the place at the microscope, performing "the readings of the height the weights were lifted . . . with excellent precision."[67] In a much stronger sense than some of the machines for psychophysiological measurements of time that were to follow, therefore, Helmholtz's research machine was not a "bachelor machine" in the trivial sense.[68]

In the mid-1860s, the Utrecht physiologist Franciscus Donders was told by his wife about a magazine report on a new instrument, the phonautograph, which soon afterward was to become the decisive attractor of his psychophysiological experiments on time. Yet while Ernestina Donders did not directly participate in her husband's research, Helmholtz opened his laboratory to his wife as secretary, assistant, and test subject, and thereby increased the performance of his experimental assemblage.[69]

Once it got going, the Königsberg research machine produced a comparatively monotonous semiotic output. It consisted essentially of numbers. The "Preliminary Report" limits itself to summary figures on the length of the frog nerves and the split seconds it took the nerve stimulus to propagate in the nerve. The "Measurements," in contrast, provide detailed minutes of individual series of experiments that contain—in keeping with the complicated calculations—a large number of variables. After the date, the distance between the two nerve points and the needle's deflection before and after the time measurement proper are listed. Following these, arranged in columns, are the numbers assigned sequentially to each "observation," the mass of the weight suspended, the height the weight was lifted, and the deflections of the galvanometer needle during the stimulation at the "more distant" and the "closer" point. This list is supplemented by the averages, obtained "according to the rules of probability theory,"[70] for the galvanometer deflections observed for the various nerve points. In addition, there is the "probable error" of these averages and of individual observations. On this basis, finally, the "time difference because of the propagation" and the "propagation speed," converted into meters per second, are listed (Figure 18).

As the laboratory logbook the research couple kept from March to June 1850 shows, the published protocols were based on a handwritten system of notations that registered the observations made on successive days (Figure 19). The formatting they adopted in the logbook, accordingly, was a simple one. Pencil lines horizontally marked off seven rectangular boxes on the blank sheets. On the vertical outside edge, each page contained an empty space that took up between a quarter and a third of the total available space. The numbers were then entered into the boxes according to a more or less unchanged schema: first the deflection of the needle before and after a given observation, then the weight put on the muscle, finally the deflection of the galvanometer needle and the height the muscle lifted the weight. At the beginning of each series of tests, the date and sequential number of the individual observation were registered right next to the boxes.

The margin also provided space for first calculations and short remarks such as "Ajusté" (adjusted); "The muscle was badly attached at the top, tore off in the end"; or "Continued with the second calf in the afternoon."[71] From time to time, such remarks were also written directly into the field of a specific experiment: "Long, irregular twitching," "Without weight," or "Adjusted as precisely as possible."[72] Additionally, some remarks in this logbook had an arc-shaped pencil mark added on the side, presumably while the results were reviewed. None of the experiments recorded here, however, found their way into a publication. The "Measurements" do mention series of experiments performed in May 1850.[73] In this regard, there is a temporal overlap between this publication and the laboratory diary. But neither the dates nor the roman numerals denoting the experiments in the "Measurements" can be correlated with individual entries in that diary.

B. Linker Muskel; Nervenstrecke 40 mm.; Ablenkung vorher 113,05, nachher 112,20, im Mittel 112,62.

No.	Ueberlastung.	Erhebungshöhe.	Differenz der Ausschläge bei Reizung der entfernteren Nervenstrecke.	Differenz der Ausschläge bei Reizung der näheren Nervenstrecke.
15	100	0,65	128,14	
16	—	0,70	133,40	
17	—	0,72		132,06
18	—	0,70		118,19
19	—	0,68	125,75	
20	—	0,68	119,80	
21	—	0,70		119,84
22	—	0,68		120,71
23	—	0,68	127,77	
24	—	0,68	133,53	
25	—	0,68		130,35
26	—	0,70		123,21
27	—	0,70	136,89	
28	—	0,75	129,28	
29	—	0,72		123,58
30	—	0,77		125,29
Mittel			129,25	124,15
Wahrscheinl. Fehler des Mittels .			± 1,15	± 1,09
Derselbe der einzelnen Beobachtung			± 3,258	± 3,097
Zeitdauer zwischen der Reizung und der Erhebung des Gewichts			0,03164	0,03039
Wahrscheinl. Fehler derselben .			± 0,00027	± 0,00026

Zeitunterschied wegen der Fortpflanzung: 0,00125 ± 0,00038

Fortpflanzungsgeschwindigkeit: 32,0 ± 9,7 Mt.

FIGURE 18. A page from Helmholtz's "Measurements" (1850). In the first column on the left is listed an experiment's sequential number, then follow the weight hung from the muscle and the height this weight was lifted by the contraction. The other columns contain the magnitudes of the needle deflections on the galvanometer. Below the table, the calculation of median values and probable error margins. Reproduced from Hermann Helmholtz, "Messungen über den zeitlichen Verlauf der Zuckung animalischer Muskeln und die Fortpflanzungsgeschwindigkeit der Reizung in den Nerven," *Archiv für Anatomie, Physiologie und wissenschaftliche Medicin* (1850), 276–364, here 342.

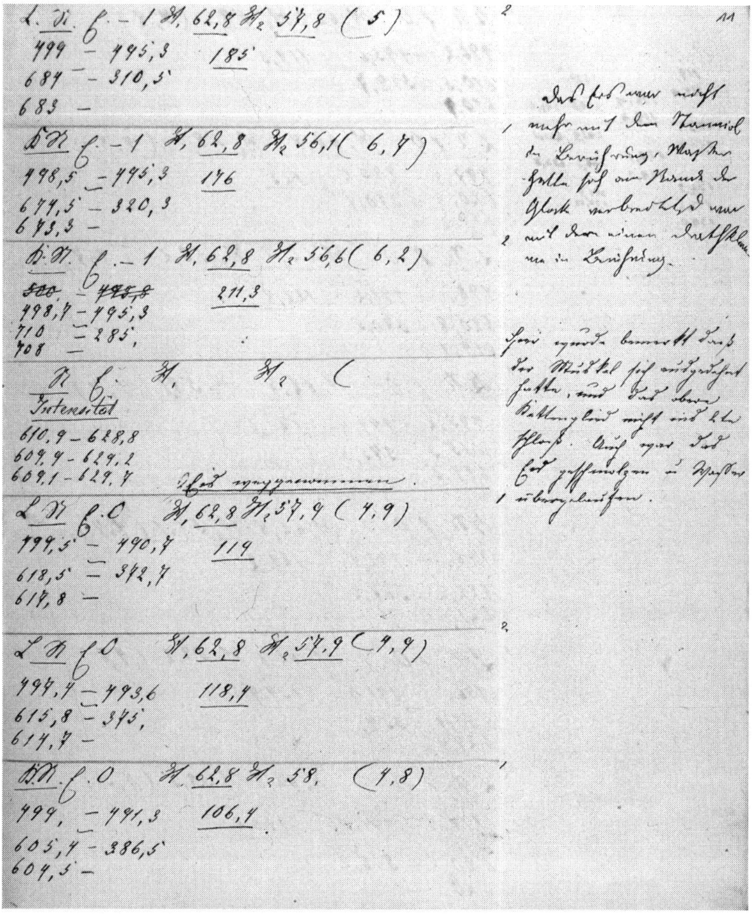

FIGURE 19. A page from the laboratory logbook Hermann and Olga Helmholtz kept from March to June 1850. The objects of this logbook are experimental time measurements on artificially cooled nerves. The left area of the page served to note down numerical values, the right area served for first calculation and remarks. On the top right, Helmholtz remarks on a series of experiments of March 30, 1850: "The ice no longer touched the tinfoil, water had spread in the belly of the bell [jar] and touched the wire tongues." Below, Olga adds: "Later noticed that the muscle had expanded and the upper chain link did not hook into the second. Also, the ice had melted and the water had spilled over." Reprinted with permission from Archiv der Berlin-Brandenburgischen Akademie der Wissenschaften, Berlin, Helmholtz Papers, NL Helmholtz 547: 11.

It is worth noting that the kind and number of variables numerically recorded during the experiments changed in the course of the study. This too, albeit on a smaller scale, indicates that Helmholtz's research took and cleared new pathways. This concerns, for example, the time the frogs were kept. In the experiments that formed the base for the "Preliminary Report," Helmholtz had experimented with animals that, as he writes retrospectively, were "weakened by four months of captivity and hunger."[74] The stimulatability of these frogs was thereby changed in such a way that at different nerve points currents of different intensities had to be used in order to produce the "same mechanical effects"—an unwelcome complication of the experiment, which was complicated enough as it was and for whose success a maximum of identity in its repetitions was decisive. The published protocols of the "Measurements" thus also list a given frog's laboratory age. Helmholtz in his new experiments furthermore tried to work as much as possible with "freshly caught frogs"[75]—which had not been possible in the winter months.

But the frogs' freshness was not the only factor whose importance was revealed as the study progressed. Heat, too, had an effect. To be sure, the "Preliminary Report" had already listed, in addition to the frog nerves' length and the propagation speed of the nerve stimulus, the average external and internal temperatures to be maintained for the keeping of the frogs and the actual experiment, respectively. This evidently reflects Helmholtz earlier work on the phenomenon of organic heat. As he explains in the "Measurements," these temperatures were indeed included because he suspected "that the propagation speed probably diminished as the temperature diminished."[76] In fact, however, he "only later"—after the experiments done in the winter of 1849–50—became aware concretely of the influence temperature exercised on the propagation speed of the nerve stimulus.

In his first experiments, Helmholtz had still "neglected determining the temperature of the room."[77] The "Preliminary Report" thus worked with retrospective estimates, which nonetheless were meant to explain the spread of time values (0.0014 to 0.0020 seconds) by the differences in temperature. The higher time values correspond to "the colder days."[78] In the "Measurements," however, temperature is not registered until the protocols of series of experiments done in May 1850. Helmholtz no doubt attached great importance to this factor as a whole, for he investigated it in special experiments. The "Measurements," for example, conclude with a report on some experiments in which the nerve was chilled in circumscribed spots with the help of small pieces of ice. The conclusion Helmholtz drew from these tests was analogous to the suspicion cited above: "In the cooled-down spot, the propagation speed of the stimulus is significantly reduced."[79]

Even in the conclusion, therefore, Helmholtz's investigation continues its drifting. After the transition from chronography to chronoscopy and from muscle action to nerve stimulation, after he had given up working all by himself and subsequently cooperated with his wife Olga, the nerve stimulation now appeared no longer as a purely temporal but also as a thermal phenomenon (on this point, compare the comments in the laboratory logbook, Figure 19).

It is probably at this point that the time measurements on the nerve link up most clearly with Helmholtz's early studies on muscular activity, which had been centered in particular on the problem of heat. In his study "On the Generation of Heat in Muscle Activity," published in 1848, Helmholtz had shown that the contractions of a muscle produce heat. He thus dealt a decisive blow to the conception that body heat comes only from the blood. In the same essay, Helmholtz had also communicated some observations "on the generation of heat in the nerves," but he had found only "vanishingly small" changes in temperature.[80]

At the end of the "Measurements," he returned to this phenomenon because he had observed a connection between time and heat. What at first seemed to be only a temporal matter within the organic body now turned out to be dependent on the ambient temperature—inside and outside the body. With respect to the isolated frog muscle one could thus say *que le temps se prolongeait ici dans le temps*—time here continued into the weather.

This would not have surprised Proust. Sensitive to dust, noise, and light, the author had done all sorts of things to shelter himself from the outside world in his apartment on the Boulevard Haussmann—almost like a sample under a bell jar. On the room's two big windows facing the street on the third floor, the shutters always remained closed. The curtains and the cotton-stuffed drapes were always drawn. Since July 1910, large corkboards lined the walls and ceiling to protect Proust from noise.[81]

Here, too, the seclusion and calm obtained in the apartment turned out to be the paradoxical precondition of a heightened perception of the outside. Not unlike his narrator, Marcel, the author now found himself in a room in which "each sound serve[d] only to make the silence apparent by displacing it,"[82] and we may assume that this precisely enabled Proust—as it did his narrator—to twitch and to jump inside, "like a machine with the brake on running in neutral."[83]

In fact, Marcel in his closed room manages to participate in the everyday events taking place around him even more intensively than before. After he has locked himself in with Albertine as his prisoner, for example, he observes the following:

> It was above all inside myself that I heard with delight a new sound struck from the inner violin. Its strings are tightened or slackened by simple variations in temperature, in exterior light. Within our being, that instrument which the uniformity of habit has reduced to silence, melody springs from these changes, these variations, which are the source of all music: the weather on particular days makes us move immediately from one note to another. . . . Only these inner changes (though they came from outside) brought the outer world alive again for me. Connecting doors, long been walled up, were opening again in my brain.[84]

Articulated in musical terms, time here detaches itself from external authorities, such as clocks, and expands into the weather, into a periodically variable phenomenon, as determinate spatially as it is temporally, that takes in the entire body and redefines this body as "time-space." The concluding remarks on the relationship of nerve activity and heat suggest that in the "Measurements," Helmholtz, too, was on the tracks of such a time-space.

FOUR

Networks of Time, Networks of Knowledge

> We are the synchronizers.
> We are electronic performers.
>
> —AIR[1]

We can also observe an interlocking of the internal with the external in the way the Königsberg research machine worked, especially in the way it co-ordinated its effects in time. Although it functioned in physiological back-waters and in the calm of a university building in East Prussia, this machine closely connected with comparatively hectic scientific and technological activity in Berlin and Paris.

Besides the publications that picked up on the work of the machine one after the other—including the "Preliminary Report" and the "Measurements"—Helmholtz used in particular his correspondence with du Bois-Reymond to keep his scientific milieu apprised of the progress of his machine work. Via du Bois-Reymond, Helmholtz reached the Physical Society and, most important, the Prussian Academy of Sciences in Berlin and the Académie des sciences in Paris. In addition, he thereby kept in con-tact with Johannes Müller, Gustav Magnus, and Alexander von Humboldt, who was still an important mediator between German- and French-speaking scientific cultures.[2] In return, du Bois-Reymond channeled the most recent developments in electrophysiological research into letters to Helmholtz, in part in informal summaries, in part in the form of enclosed publications.

By 1848, before Helmholtz had moved to Königsberg, du Bois-Reymond had already published the first volume of his *Studies on Animal Electricity* and dedicated it to Johannes Müller. One year later, he published the first part of the second volume and dedicated it to Alexander von Humboldt. In 1850, while Helmholtz in Königsberg set about writing down the "Measurements," du Bois-Reymond was busy editing the second part of the second volume of his *Studies*. At this time in particular, the letters exchanged between the two physiologists go into great scientific and technical detail.

This could be called cooperation at a distance, with protuberances of local research structures serving as connections between laboratory practices in central and eastern Prussia, almost the organization of a single scientific project distributed across people and spaces. In fact, du Bois-Reymond wrote to Helmholtz around this time: "We're now working properly hand in hand, and that's how something may come of it, since the work is too much for one person alone."[3] And, writing to Helmholtz about parallel phrasing that showed up in certain texts published by both without their having discussed them, he even speculates that these parallels are due to a "conformity in the movements of our brain molecules."[4]

In any case, the discourse of interlocking, of working hand in hand, was not meant only metaphorically. The two scientists were indeed concerned with more than exchanging and coming to agreement on physiological contents and concepts. Thus Helmholtz kept in touch, via du Bois-Reymond, with Berlin mechanics, although he had most of the technical components of his experimental setups fabricated in Königsberg.[5] In Berlin, these technicians included, for example, the precision mechanic and mirror cutter Johann Oertling and the electrician Johann Georg Halske, in whose Telegraph Construction Works (which he ran with Werner Siemens) du Bois-Reymond had his galvanometers built and where he sometimes built them himself.[6]

Via letters to du Bois-Reymond Helmholtz also conducted his negotiations with Moritz Veit and other Berlin publishers, no matter whether the issue was corrections in the manuscript of the "Measurements," changes in the drawings to go with that text (should the contacts in the frog frame's suspended steel frame be shown as open or closed?), or the right engraver for converting the drawings for print. Even the question of what paper to print the corresponding plate on was a matter of epistolary discussion between Helmholtz and du Bois-Reymond.[7]

Besides public and private writings—publications and correspondence—Helmholtz's research machine interconnected with the centers of scientific and technological activity farther west on the level of material culture as well. Within the laboratory at Königsberg University, this machine was a network-like arrangement of heterogeneous components. Within Prussia or Europe, even, it became the nodal point of a network of much larger dimensions, a network of very different projects in science and technology.

For these projects, the connection with the machine in Königsberg was established primarily by Helmholtz's taking recourse to the means Pouillet had suggested for measuring extremely short portions of time. In choosing this recourse, Helmholtz linked up with a polemic, as lively as it was sweeping, which scientists, instrument makers, and military officials had been engaged in since the mid-1840s. This was a debate about applying electromagnetism to the problem of precisely determining and communicating time. It was shaped by experiences in the construction of telegraphs and the expectation of introducing and spreading appropriate devices for measuring time in ballistics—that is, the interest in selling them in large quantities to the military.

The debate had been set in motion by the lecture of December 23, 1844, in which Pouillet had presented his electromagnetism-based method of measuring time to the Paris Academy.[8] Only a few days after Pouillet's lecture, on January 20, 1845, the physicist, clockmaker, and telegraph builder Louis François Clément Breguet presented the same Academy with a device for precision measurements of time.[9] This device, which Breguet developed with a Russian military official by the name of Konstantin I. Konstantinov, marked time on a rotating cylinder with the help of an electromagnetically controlled stylus. According to Breguet, it was capable of "measuring the velocity of a projectile at different points of its trajectory." In a way similar to Pouillet, who had compared his method to a ballistic pendulum (see chapter 3), Breguet, too, established a connection with military applications. At the same time he emphasized that he had had the idea for an application of electricity to the measurement of small amounts of time significantly earlier than Pouillet. At least he claimed that the construction of the device he now presented had taken place "a year" earlier.[10]

Just four months later, the English physicist and instrument maker Charles Wheatstone reacted to Breguet's presentation in yet another lecture before the Académie des sciences. There, on May 26, 1845, Wheatstone claimed to have already built such an instrument "for the purpose of measuring rapid motions, and especially the velocity of projectiles" at the beginning of 1840.[11] Since then, he claimed, he had constructed a whole series of such instruments that were all to be understood as "derivations" from the electrical telegraph he had constructed with William Cooke in 1837.

Wheatstone's claim was emphasized by his introduction into the debate of a term used neither by Breguet nor by Pouillet: "chronoscope," from the Greek *khrónos*, time, and *skopein*, to contemplate or see. The term was already in use, synonymous with "chronometer," in the eighteenth century. In Wheatstone, however, it functioned as a new collective name for electromagnetically controlled clocks and precision chronometers. It also linked up with a series of similar expressions the English physicist had introduced for other instruments of his: the "rheoscope," for example, and the "electroscope" (both devices for measuring electrical currents); the "gyroscope" (an instrument for showing spinning movements and for proving the earth's rotation around its axis); the "pseudoscope" (for reversing optical relief effects); and, not least of all, the famous "stereoscope." The electromagnetically controlled clocks thus took quarters in a veritable arsenal of scopes.

After this introduction, Wheatstone revealed that he, too, knew Konstantinov. He had met him in London as early as 1842 and, not long afterward, had built an "electromagnetic chronoscope" for him. In January 1843, he sent this device, Wheatstone said, to Paris, where the Russian official was by then staying. The message was clear: Wheatstone insinuated that, with Konstantinov's help, Breguet's construction of his time-measuring device was inspired (to put it mildly) by his, Wheatstone's, invention, or (to put it less mildly), that Breguet's device was a rip-off. The English physicist accused Breguet of this plagiarism and furthermore criticized the apparatus in strong words:

> Had M. Breguet been better informed on the means by which I would obtain a series of successive measurements corresponding to a same path, he would have found that what he proposes to obtain, even with a dozen electro-magnets, would have been obtained in a much more efficacious manner by means of one alone.[12]

For Wheatstone, the device Breguet had constructed for measuring very short amounts of time was "much less exact, much more complicated, and more costly" than all the chronoscopes he had designed and built since 1840. Wheatstone, however, did not submit concrete evidence for these allegations, not least because he—like those who had spoken before him—could not supplement the published version of his lecture in the *Comptes rendus* with images. He was left with assuring, after the fact, that he had submitted drawings of his chronoscopes to some members of the Paris Academy as early as 1841 and that he had even given some of them to Pouillet for copying.

Despite a series of historical studies dealing with electromagnetically controlled devices for measuring short time spans in the nineteenth century, it has so far not been possible to pinpoint the exact beginning of their lineage.[13] Most likely, there isn't any *one* point at which "the" chronoscope was invented. According to everything we know, no patent was ever granted. And by 1845, by the time Breguet and Wheatstone made their appearances before the Paris Academy, it was too late for a beginning in this sense. By the mid-1840s, the application of electromagnetism to the problem of time was anything but a novelty. As the example of Gauß and Weber shows (see chapter 3), this application suggested itself as soon as one spoke of electrical telegraphy. In 1839, the physicist and telegraphy pioneer Carl August Steinheil, who kept up a correspondence with Gauß and Weber in Göttingen, had a system of "galvanic clocks" patented. In 1840, Wheatstone had presented an electromagnetically controlled clock in the Royal

Society's "apartments," and Alexander Bain's work on networks of so-called electrical clocks goes back to that same time (Figure 20).[14]

Wheatstone's remark that the chronoscope was to be understood as a derivation from his work with telegraphs was obviously part of his attempt to secure a claim to priority. But this same remark also meant that Wheatstone laid claim to a derivation from a technology that was itself nothing but a derivation. And because it was only a derivation, other scientists and engineers were perfectly at liberty to make their own derivations from electromagnetism in order to advance, each in his own way, the networks of

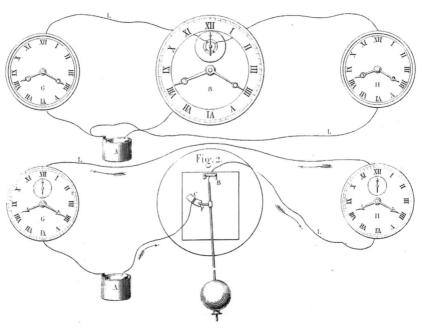

FIGURE 20. Illustration of a system of electrical clocks (1852). At the center, the pendulum clock that emits the signals, on the left and right the dials and hands it controlled. (A) denotes the source of energy, a galvanic element. Below, the view of the back of the entire system shows how the time signal is given by a pendulum closing and opening a contact. Reproduced from François Moigno, *Traité de télégraphie électrique, comprenant son histoire, sa théorie, ses appareils, sa pratique, son avenir, sa législation*, 2nd ed. (Paris: Franck, 1852), table 19.

communication and of time then emerging—and to advance them not merely in technological terms.

The polemic about time, telegraphy, and speed quickly spread. In the fall of 1845, it was taken up and continued in Berlin before the Physical Society, which du Bois-Reymond had cofounded and of which Helmholtz was one of the first members admitted. On October 3, 1845, Werner Siemens, at that time a lieutenant in the Prussian artillery, addressed the Society with a lecture, "On the Application of Electrical Sparks in the Measurement of Velocities."[15] For Siemens, the debate taking place in the Paris Academy was in fact nothing but a "priority dispute," which he did not hesitate to join.

Siemens claimed that it had been the Artillery Inspection Commission at Berlin that had been the first to develop an electromagnetically controlled precision chronometer. In 1839—one year before Wheatstone—this commission had asked the Berlin "clockmaker Mr. Leonhardt" to build a chronometer that was "suitable for measuring very small parts of time and could be engaged and arrested by means of magnetic force."[16] A short time later, Leonhardt—who, like Siemens, was an early member of the Physical Society—was commissioned by the General Staff of the Prussian army to run a series of experiments on electrical telegraphy, which Siemens also participated in.[17]

At the same time, Lt. Siemens criticized the time-measuring instruments described by Wheatstone *and* Breguet and made suggestions for improvements based on his own experiences with the instrument. What seemed particularly problematic to him was the fact that both Wheatstone and Breguet left the recording of the "time marks" to electromagnetically controlled "anchors" on the surface of a rotating cylinder. Such anchors, he writes, "produce significant friction in the moment of the impact on the cylinder, which detrimentally affects its [the cylinder's] uniform movement."[18] In addition, the "considerable friction" in the cylinder's screw thread contributed to disturbing the evenness of the rotational speeds—a criticism whose basic elements Helmholtz was to adopt in the "Measurements" to justify his abandonment of the curve method. As we have already seen, friction was the main reason the physiologist, too, thought this graphical method could not yield precise measurements of time.

The improvement strategy Siemens had employed in his work on the Inspection Commission's time-measuring apparatus since 1842 also fore-shadowed certain aspects of the future development of Helmholtz's research machine. According to his statement before the Physical Society, Siemens's main goal had been "to do away with every mechanical interme-diary element between the bullet"—the projectile whose velocity was to be measured—"and the time indicator."[19] The "electrical chronoscope" he developed, to be sure, still relied on a rotating cylinder. But instead of being recorded with a drawing stylus, the time markings on the cylinder were now done by a spark, an electrical discharge. The stylus thus no lon-ger touched the cylinder. There was no friction.

To Helmholtz, this solution was convincing all around—or so he says at the end 1850 in a popular lecture, "On the Methods of Measuring Very Small Portions of Time, and Their Application to Physiological Purposes." In this lecture, Helmholtz reviewed large parts of Siemens's report on the priority debate, both in his presentation and in a subsequent publication in the *Advancements in Physics*, published by the Physical Society.[20] At the same time he emphasized that Siemens had had "the happy idea of doing away with all the mechanical interventions, and of permitting the electric-ity itself to mark the times."[21]

The research machine at the basis of the time measurements Helmholtz described in the "Preliminary Report" and the "Measurements" had, of course, abandoned the rotating cylinder completely. Yet the galvanometer method employed instead did accord with Siemens's chronoscope in one respect. It, too, allowed the actual process of measuring time to take place almost without "mechanical interventions," without friction; it was essen-tially an electromagnetic process, the deflection of a galvanometer needle suspended from silk threads above a coil (see Figure 14).

In a certain way, then, electricity "itself" was "marking" time here as well—even if it did not leave any permanent traces. By referring to Pouil-let's method, therefore, Helmholtz did not only go back to the beginning of a polemic about electromagnetically based procedures for measuring time: Beyond that, his explanation of how his research machine functioned made use of the very arguments made during the course of that debate and its shift from Paris to Berlin.

The Italian Problem

Grafted onto this connection with the material culture of telegraphy and precision time measurement as it developed in Paris and Berlin were the connections Helmholtz's research machine entertained with its scientific context more narrowly defined. Given the intense exchange between science, industry, and military in the milieu of the Physical Society, it is hardly surprising that Helmholtz was neither the only nor even the first organic physicist to be interested in the new methods of precision time measurement—especially since the problem was by no means a new one in Berlin's physiology circles.

As early as 1837, Johannes Müller, who had taught both Helmholtz and du Bois-Reymond at the university, had commented on the problem in his textbook, *Elements of Physiology.* In a remarkably sober way, he opened the section discussing the "mode of propagation of nervous action in the different nerves" with the assessment that physiologists "shall probably never attain the power of measuring the velocity of nervous action."[22]

In the background of this assessment stood a contrast with the procedures astronomers had developed for determining the speed of light, comparing observations of the movements of Jupiter moons with the positions to be expected thanks to calculation. Whereas this procedure was based on a "comparison between immense distances," the spatial realities of the living body precluded a correlate procedure.[23]

Yet seven years later, new means for determining velocities at, clearly, a much higher level of precision had become available—and they had become the crystal nucleus of a highly publicized polemic in Paris and Berlin that was difficult to ignore.

Du Bois-Reymond, not Helmholtz, was the first to react in this situation. In yet another lecture before the Physical Society, du Bois-Reymond, on March 7, 1845, directed attention to the new methods of measuring short periods of time and offered some initial thoughts on their application in physiological research. This means that the author of the *Studies* suggested investigating muscle and nerve activity using the procedure already outlined by Pouillet at a time when the polemic about chronoscopy, though underway, had not yet reached its climax. A publication to this effect in the

Society's journal, the *Advancements in Physics*, however, did not occur. The *Advancements* did not mention the lecture until three years later, in a most abbreviated and not exactly informative one-liner: "Method for measuring the velocity of muscle and nerve activity."[24]

The French public, in turn, was informed about the project outlined by du Bois-Reymond in more detail by 1846. In an overview of advancements in the physical sciences outside France, the *Revue scientifique et industrielle* ran a short paragraph on the Berlin physiologist's lecture:

> Monsieur du Bois-Reymond presented the project of a method that would serve to determine by experiment the propagation speed of the nervous principle and that of muscle actions. This method rests essentially on the principle indicated by Monsieur Pouillet for measuring extremely short time spans.[25]

An additional sentence concretizes the procedure: "All that needs to be done is to make sure that the current is interrupted by the effect of and in the same instant as the contraction excited by the establishment of that current."[26]

This was, in one sentence, a fundamental characteristic of the machine Helmholtz was to work with four years later in Königsberg, in both of the fields of application named by du Bois-Reymond: the physiology of muscles as well as that of nerves. Yet the problems raised by the permanent closing of the time-measuring circuit could of course not be foreseen in 1845.

But du Bois-Reymond and Helmholtz were not alone. They were not the only physiologists who, in the wake of the confrontation between Pouillet, Breguet, Wheatstone, and Siemens, began to be interested in applying electromagnetic procedures to measuring short time spans in the physiological laboratory. In 1847, the Italian physicist and telegraphy expert Carlo Matteucci, who had worked in electrophysiology since the 1830s, also turned to the new chronoscopes and chronographs.[27]

Matteucci taught and conducted his research at the University of Pisa, yet he had studied in Paris and had scientific contacts in France, Germany, and England. He corresponded with, among others, François Arago in Paris, Alexander von Humboldt and Johannes Müller in Berlin, and Michael Faraday in London. Matteucci published a large share of his research in French, partly as monographs—such as his 1844 *Traité des phénomènes électro-physiologiques des animaux*—but above all as journal articles. In the years between 1845 and 1850 alone, about fifty were published, and in the course of his entire career, more than two hundred and fifty.

In quick succession, Matteucci reported on new "experiments" and "investigations," but there were also letters to Arago or von Humboldt that appeared in the Academy of Science's *Comptes rendus*, followed by translations into German, published especially in the *Notizen auf dem Gebiet der Natur- und Heilkunde,* and into English, for example in the *Electrical Magazine* or the *American Journal of Science and Arts.*

After the Royal Society in London had honored him in 1844 with the Copley Medal for his contributions to electrophysiology, the prestigious *Philosophical Transactions* published several "series" of his "Electrophysiological Researches" at short intervals. These series testify to his interest in the new possibilities precision time measurement opened up for physiological research.

It was at the Royal Society that in June 1847, about two years after du Bois-Reymond's lecture before the Berlin Physical Society, Matteucci presented, as preliminary apex and conclusion of his "Researches," a number of experiments that aimed at subjecting the "duration of these [frog muscles'] contractions" to precise measurement.[28] Similar to what Helmholtz was to do, Matteucci worked with suspended muscle samples taken from frogs, which he weighed down and stimulated with an electrical current, and, just as Helmholtz did, he picked up on the procedures presented and discussed in the priority debate concerning the electromagnetic measurement of time.

Matteucci explained, in a somewhat circumlocutory way at first, that the "chronometer" he used in his experiment was a device "which appears to me analogous in principle to that upon which the instruments invented by Messrs. Wheatstone and Breguet for measuring the velocity of projectiles are founded."[29] Yet, as he makes clear as he proceeds with his presentation, he had ordered this time-measuring instrument from none other than Breguet in Paris. All other parts of his experimental setup, like the frog frame, also came from the workshop of Breguet, with whom he had already worked for a few years.

The time-measurement experiments the Italian physicist performed with the equipment thus obtained concerned two aspects of muscle movement. In the first series of experiments, Matteucci studied the interval between the contractions to be observed in the frog's calf muscle during repeated punctual stimulations. He determined the time elapsed by way of the contracting muscle opening and closing contacts and thereby opening and closing an electrical circuit that in turn caused a magnet to move such that an anchor "at each movement strikes against the knob of a chronometer."[30]

Matteucci thus did *not* make use of Pouillet's method, and he tried to measure the time passing *between* two contractions, not the time of the muscle contraction itself, never mind that of the nerve stimulation. And yet the basic principle of his experimental setup can be seen as putting into action the plan sketched by du Bois-Reymond. That project had suggested interrupting the current precisely at the moment of the contraction, which had, in turn, been caused by a circuit being closed. Yet Matteucci did not say much about how his chronometer functioned. From what we can tell, it seems to have consisted of a mechanical clockwork set into motion and stopped again by means of electromagnetism (Figure 21). This was *one* version of the chronoscope developed by Wheatstone and represented a functional principle that was widely applied in 1847.

In the case of the instrument he had developed together with Konstantinov, however, Breguet had been guided by a different version of Wheatstone's chronoscope, and he used a rotating cylinder for recording curves. In the first of Matteucci's series of experiments, this was obviously not the case. This might be one reason why the level of precision in the results he obtained was rather low. The Italian researcher was aware of that and merely reported that, as time went on, the interval between two muscle contractions became greater. Within a space of ten to fifteen seconds, it went first from 0.25 to 0.33 seconds, then to 0.41 seconds, and finally to 0.58 seconds[31]—what would later be referred to as a *fatigue* effect.

In the second series of experiments, Matteucci took on the task of measuring the two phases of muscle movement: the contraction proper, the weight's upward movement, and the subsequent dilatation, the weight's downward movement. Without presenting any reasons, he replaced here

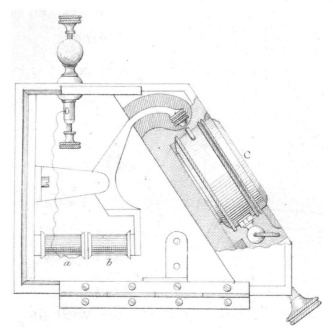

FIGURE 21. Illustration of the Breguet chronometer used by Matteucci in his time-measuring experiments on suspended frog samples (1847). An electromagnet (a, b) controls a curved arm or anchor that sets the chronometer (c) into motion and stops it again. The chronometer is seen from above. Reproduced from Carlo Matteucci, "Electro-Physiological Researches, Seventh and Last Series Upon the Relation between the Intensity of the Electric Current, and That of the Corresponding Physiological Effect," *Philosophical Transactions of the Royal Society London* 137 (1847): 243–48, table.

the electromagnetic method he had previously used with a different procedure. He now recorded curves, a procedure that had also played a part in the device constructed by Breguet and Konstantinov.

As Matteucci explained, once again in a somewhat circumlocutory way, this was "the same method which, I believe, the celebrated Watt adopted in the first instance for determining the velocity of the pistons in his machines."[32] In the same year, 1847, in which Ludwig introduced the kymograph to physiological laboratory research (without, however, referring to Watt in doing so),[33] and more than one year before the frog drawing machine Helmholtz was to construct in Potsdam, Matteucci appears to have put a time-sensitive curve drawing procedure to practical use in (muscle) physiological research, a procedure explicitly inspired by the indicator diagrams used in stationary and mobile steam engines.

As becomes clear from his short and, unfortunately, not illustrated description of the setup, Matteucci did *not*, however, use a rotating cylinder: "A fine point was attached to the little shank fastened to the leg of the frog, which point scraped, during the contraction, against a rapidly revolving smoked disc, of which the rotations were perfectly uniform."[34] Matteucci adds that since the exact time the disk needed for a single rotation was known, the trace left on the round disk by the point during the muscle's contraction and subsequent dilatation made it possible to determine the length of the contraction and the dilatation.

The result he had obtained this way "in a great variety of experiments,"[35] however, consisted only in a summary statement. The duration of the contraction, he writes, is significantly shorter than that of the dilatation, and this difference increases if the muscle is used for a longer time. Like Helmholtz after him, therefore, Matteucci was already unable to take precise time measurements with the graphical method. In conclusion, he merely gives an estimate of how long the "real contraction," without dilatation, takes: "less than 1/100th of a second."[36]

Du Bois-Reymond, with Helmholtz, against Matteucci

Despite the fragmentary character of Matteucci's report as a whole, despite the obscurity surrounding the exact functioning of the Breguet chronometer he used, and despite the lack of precision in the results he obtained in his time measurements, there could be no doubt that the Italian physicist had taken first steps toward making the project first outlined by du Bois-Reymond a reality: the project of applying the newer methods of measuring time to problems in the physiology of muscles and nerves. More than that, in making use of both electromagnetism *and* the graphical method, Matteucci staked out the technological poles between which Helmholtz's research machine, too, was to move: chronoscopy on the one hand, chronography on the other.

Physiologists in Berlin would have been interested but also rather disturbed to learn this. Du Bois-Reymond, for one, must have experienced the fact that the initiative came from Matteucci, of all people, as an unpleasant *déjà vécu*. Ever since he had turned to electrophysiology in the early 1840s, he had time and again made reference to the work of the Italian physicist—yet his attitude was anything but uncritical. In 1841, Johannes Müller had handed the young du Bois-Reymond a copy of Matteucci's recently published *Essai sur les phénomènes électriques des animaux*, "asking me to repeat the experiments on frog electricity it contained and, where possible, to develop them further."[37]

The initial orientation of his research was accordingly rather close to Matteucci's. Du Bois-Reymond had picked up on the questions raised by Matteucci, seven years his senior, and let himself be guided by the methodology of his Pisa colleague. At the same time, however, he sought to work more carefully and more coherently—both theoretically and practically.[38]

The practical concerns included technical concerns—above all the construction and employment of the galvanometer—but also touched on the question of writing and publishing. In 1843, du Bois-Reymond had published in the *Annalen der Physik und Chemie* a "Preliminary Summary" of the first results of his electrophysiological studies. Convinced of its significance, he translated the text into French and sent it, via Alexander von Humboldt, to the Académie des sciences for publication in the *Comptes rendus*. But publication was more than delayed; it never took place. Even worse, two years later, the *Comptes rendus* contained one of Matteucci's "Letters" to von Humboldt, in which the Italian claimed for himself the establishment of fundamental facts that had already been described in du Bois-Reymond's essay—first, that there is no particular "frog electricity" but only one kind of electricity, and second, that the muscles of organic individuals can produce electricity.[39]

Now Matteucci had claimed the discovery of these facts for himself. Such at least was the impression du Bois-Reymond was under shortly after the letter appeared in the *Comptes rendus*. Incensed, the Berlin physiologist in May 1845 drafted a letter to von Humboldt as well, demanding that the

Paris Academy's journal publish a rectification: "I regret having to take this step, but I must not fail to claim the uncontested right of priority I believe to possess on this point"—but "this point" is not explained.[40] This may have been *one* of the reasons why du Bois-Reymond in the end did not send the letter to von Humboldt: The fact at issue tended to disappear in the heat of the battle. Other reasons become clear from the rest of the unsent letter. For du Bois-Reymond, the question of priority was part of a much larger problem that had to do, on the one hand, with the journal culture that already existed in his time but expanded ever more rapidly and, on the other, with fundamental differences in how one conceived of experimentation.

It is probable that the Berlin physiologist's view, aimed as it was at comprehensive questions, had been schooled in the debate that raged in England and Germany in the 1830s about who should be considered the discoverer of the physiological mechanism of reflex action: Marshall Hall or Georg Prochaska. This debate had also involved the priority of Hall over Johannes Müller, du Bois-Reymond's teacher. Already in that debate, publication formats had played an important role: Hall had published most of his contributions in the form of articles in, among other journals, the *Philosophical Transactions*, which was later to be Matteucci's forum as well. Müller, in turn, had embedded his observations and insights in the comprehensive descriptions of his textbook, the *Elements of Physiology*. In 1833, he had to admit in that book that in the matter of the reflex mechanism, it was indeed Hall who "has, therefore, the priority"—if only by a few weeks.[41]

Some ten years later, this pattern seemed to reassert itself in the domain of electrophysiology. For while Matteucci published reports on his research at short intervals, reports that were crisscrossed by references to his own writings, du Bois-Reymond aimed for a fundamental, comprehensive and, if possible, conclusive presentation that was to contain not only the newest scientific results but also the corresponding electrophysiological research developed by other experimenters. The Pisa physicist published in Italian, French, German, and English journals, and his book publications were based on an "organized collection" of the appropriate articles, which were reviewed and, where necessary, corrected and revised—albeit without keeping an editorial record of the changes that were made.[42]

The Berlin physiologist, by contrast, did not want to translate his on-going research into a quick succession of publications that could easily be lost track of. He therefore preferred publishing a coherent book in German—even if it became necessary to do so in several installments. In fact, the last part of the *Studies on Animal Electricity* was to be published in 1884, more than thirty-five years after the first volume had appeared. This is the tenor of du Bois-Reymond's remarks to Helmholtz as well on sustained and sustainable experimentation and publication, as, for example, in a request he makes in the spring of 1850 to please stick to the "train of experiments" they had begun.[43] In light of Helmholtz's later complaints that his own work continued to "lead him further away" from the research goals he had "originarily" pursued, such a request does not seem at all unreasonable.[44]

This kind of difference in publication and research strategies is already apparent in du Bois-Reymond's 1845 draft of a protest letter. Obviously, these differences have to do with the questions of acknowledgement, appreciation, and reward that come to play an increasingly important role in the modern system of science. They refer also, however, to the relationship of science to time and to history. Without at first concretely pointing to Matteucci, du Bois-Reymond in his draft letter criticized what he saw as the excesses of the scientific journal industry. In stark words, he declared his "contemptuous" attitude toward the "kind of premature publication" practice that "is so widespread these days, which overwhelms the journals with single facts and observations without a shared connection as much as with precipitous claims to possession."[45] He himself, he declared, did not want to turn to the public until he were able "to produce a whole"—a whole capable of making its mark in his discipline, "both by the number of the new facts accumulated there and by the intimate connections that would reciprocally link them together as well as by the conclusions to be drawn from their connection."[46] This referred to the work he had begun on his multivolume work, the *Studies*.

Regarding his conflict with Matteucci, du Bois-Reymond polemicized more concretely against the lack of a sense of history in scientific research. To his mind, the main reasons for the absence of personal references and thus of acknowledgement of authorship were not "authorial vanity" or "vaingloriousness" but rather the increasing detachment of the natural sciences from their own history and, correlatively, the increasing non-historicity of scientific publications.[47] After naming Matteucci, du Bois-Reymond continued to argue in this vein against the "tendency of some physicists to completely neglect the study of writings that preceded their own research and never to mention the names of those to whom, on occasion, they owe the basic idea of their work."[48]

This aimed straight at the proliferation of publications by his colleague and competitor from Italy, but it also referred to a problem du Bois-Reymond confronted in his capacity as founding member of the Berlin Physical Society. The Society had been founded in January 1845; the first volume of the journal it published, the *Advancements in Physics*, appeared in July 1846. This periodical, designed to be a yearbook, explicitly aimed at countering the "lack of a history of literature in physics."[49] The emergence of "periodical journals," in particular, had created a situation in which "single observations, smaller essays, and works that do not stem from members of the learned societies" could be "publicized quickly and in detail." This is what Matteucci's numerous essay publications may have exemplified. As welcome as this development might be in physics in principle, the "unavoidable accumulation of material" was problematic.[50]

In the "Preliminary Report" to the first volume of the new periodical, the editor of the *Advancements*, Gustav Karsten, summarizes this dilemma as follows: "The literature of physics threatened to become too much for physicists."[51] Other journals, yearbooks, repertories, and even dictionaries had of course tried to react to this situation but so far, Karsten wrote, they had not been all too successful. For what was desirable were not just reports on "results" but also an account of the "development of studies" that had been done and presented within a given time frame, in the best case in an entire year. The *Advancements* thus aimed at being more than merely a "literary referencer." It aimed, instead, for thematically organized annual reports that "quickly and comprehensively familiarize the interested public with the progress" made.[52] Put differently, this procedure was to preclude the possibility that there would be a physicist who could completely neglect reading the works that preceded his own study—at least within the German-speaking world.

General declarations of intent and unsent letters, however, did not impede Matteucci's continuing research and publication activity. In 1850, after the first two parts of the *Studies on Animal Electricity* had been published, du Bois-Reymond therefore published, in the *Advancements*, what he himself understood to be a "sharp polemic" against the disliked investigator, the "frivolous" man of "untiring prolificacy" in Pisa.[53] It is worth noting that du Bois-Reymond's polemic also discussed the time experiments Matteucci had reported on in the *Philosophical Transactions* in 1847. The focus here, however, was *not* the procedures deployed by the Italian physicist and physiologist for the purposes of measuring time. Accordingly, the central question was not whether he had been inspired by du Bois-Reymond's early indication of a corresponding investigative technique. Instead, the criticism focused on the theoretical foundations of Matteucci's experiments and, above all, on the imprecise results these experiments had yielded. Du Bois-Reymond did not mince his words. For him, "Mr. Matteucci's numbers" were "worth just about as much . . . as if he had improvised them." That was not all. On the whole, the particular report in the *Philosophical Transactions*, and all the earlier reports, were a "mystification" such as had "not yet occurred in science."[54]

Five years earlier, in May 1845, du Bois-Reymond had not yet dared go quite that far. He never did send that letter to von Humboldt and thus never insisted on a rectification in the *Comptes rendus*. On the one hand, his "Summary" had never been published in French, which put him in a defensive position from the outset; on the other hand, he had to confront the fundamental question whether he, a young academic with just one published article, could effectively take on an article that not only was written by an undoubtedly much better established colleague but that also formed part of a whole series of such publications. Du Bois-Reymond obviously answered this question in the negative. Instead, he concentrated on the long-term work on his book that, as a "whole," was to leave a mark that was in turn to have a long-term effect—also and especially in contrast with continual publications in journals. There was no reason, however, not to speak up in essays and messages once the book was completed and his own status secured.

As for von Humboldt, du Bois-Reymond did not have to take recourse to writing to present his view of the matter; he could also talk to him in person. Von Humboldt, in any case, was soon fully aware of what to think of Matteucci even without having received the letter drafted in 1845. As the senior scholar, who was friend to both, put it in the late 1840s, the Pisa physicist had become du Bois-Reymond's "favorite enemy." Matteucci's writings had the same effect on the younger colleague as "the red cloth has on the noble bull."[55]

By comparison, the correspondence between Berlin and Königsberg was straightforward. In the letters exchanged between du Bois-Reymond and Helmholtz, we find repeated references to the "rascal Matteucci," on whom the former wishes cholera and other diseases such that "he clear the field of his presence." Claude Bernard, too, is described in these letters as a "rascal," and as, in Helmholtz's opinion, a "paltry fellow."[56] The formulations Helmholtz uses in his publications are more circumspect. In 1852, for example, he writes that Matteucci had undertaken "a great number of laborious experiments with commendable diligence." He adds, however, that Matteucci's theoretical ideas were "full of contradictions and irresolvable confusion."[57]

Du Bois-Reymond, therefore, had not overlooked Matteucci's report in the *Philosophical Transactions* of early June 1847 on applying the new methods of precision time measurement to questions in the physiology of muscles and nerves—a report that, once again, did not mention the one to whom, in this case as well, Matteucci might owe his "basic idea." Perhaps du Bois-Reymond even mentioned the concluding part of the "Electro-physiological Researches," in which Matteucci described his chronoscopic and chronographic experiments on muscle contraction, to Helmholtz. Helmholtz, in any event, refers to a conversation about the Pisa colleague's publications when he writes to du Bois-Reymond at the end of July 1847: "I would like it if you could get the essay by Matteucci for me."[58]

It is no longer possible to find out which essay is meant here and why Helmholtz did not have access to it. All that is certain is that about a year later, Helmholtz had assembled a frog drawing machine of his own and, some months later, had moved on from the curve method initially used to the electromagnetic procedure of precision time measurement—going, as it were, in the opposite direction of the path Matteucci had traveled.

From Wheatstone via Hipp to Proust

While the Königsberg physiologist laid the tracks of his own research, networks of precision time measurement and telegraphy developed in their own way—in Prussia, in France, and elsewhere. One instrument in particular was to have a decisive impact on the history of experimental physiology and psychology. Presented to the German-speaking public in 1848 by Matthäus Hipp, a mechanician and clockmaker from Reutlingen near Stuttgart, this instrument was another chronoscope inspired by Wheatstone's construction that could measure short intervals with a precision of one thousandth of a second. Like the English physicist, Hipp imagined this new time-viewer being used by the military. He accordingly presented his electromechanically controlled clock as an instrument that could be employed "in experiments on the velocity of shotgun bullets."[59]

Initially, however, it was mainly in physics classes that Hipp's chronoscope was used. In the 1850s, physicists such as Friedrich Reusch and Wilhelm Eisenlohr used it to demonstrate the laws governing the fall of solid bodies. But when Wilhelm Wundt published the first great textbook of physiological psychology in 1874, he revisited the chronoscope and recommended it for use in physiological and psychological laboratories, especially in experiments on reaction times, since reading off the dial "immediately tells the absolute time" (Figure 22).[60] And indeed, in the decades that followed, Hipp's chronoscope was introduced in practically all physiology and psychology labs, thereby becoming emblematic of the name first reintroduced by Wheatstone. In 1902, Hipp's successor, the company Peyer & Farvager, prided itself on having sold such instruments to a total of more than sixty-five research institutions in Europe and the United States, including laboratories in Philadelphia, Strasbourg, Basel, Madrid, Berlin, Zurich, Vienna, Turin, Paris, Göttingen, and Moscow.[61]

In 1845, Wheatstone had described his chronoscope as a "derivation" from the telegraph. In Hipp's case, it was the reverse: it was the work on electromagnetically controlled precision clocks that led him to telegraphy. In 1851, three years after premiering his chronoscope, he presented a writing telegraph constructed "on the American model" (Figure 23).[62] This machine was a copying telegraph that could transmit twenty-four letters.

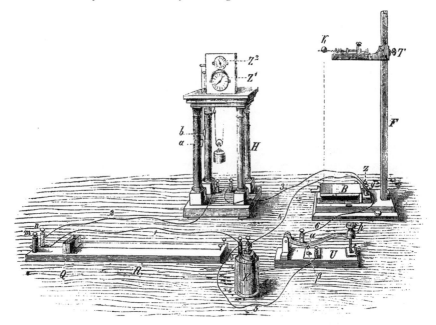

FIGURE 22. Reaction time experiment with chronoscope according to Hipp (1874). In the early 1860s, Adolphe Hirsch, director of the Neuchâtel observatory and chief of the time service, devised this setup. It consisted of a Hipp chronoscope (H), a falling apparatus (F), a telegraphing key (U), a galvanic element (K) and a rheochord (R). The test subject was asked to react to the sound made by the steel ball (k) falling onto the base of the falling apparatus. Underneath the board (B), there was an electric contact such that the mechanical pressure on the board set the chronometer into motion. The test subject reacted by letting go the telegraph key he or she had held pressed down. This stopped the chronoscope. Wundt and other psychologists later called the time measured "simple reaction time." Reproduced from Wilhelm Wundt, *Grundzüge der physiologischen Psychologie* (Leipzig: Engelmann, 1874), 770.

In its simplest form, it consisted of two parts, the "sign-emitting (the cause) and the writing part (the effect)." On the two drums, precisely coordinated in time, the letters were first recorded by way of keystrokes (transmission) and then reproduced by a writer (reception). In the functioning of the telegraph, an "extraordinary uniformity of the clockwork's operation both in the writing and the displaying apparatus" played a decisive part. It is not

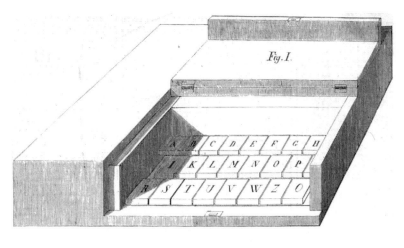

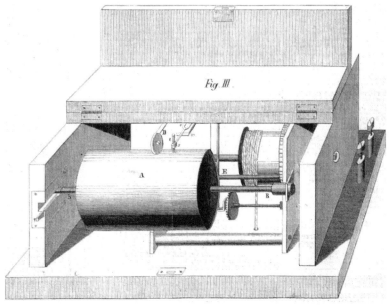

FIGURE 23. Two views of Hipp's writing telegraph developed on the "American model" (1852). On top, a keyboard for transmitting twenty-four letters; on bottom, a recording drum driven by the chronoscope mechanism. The "extraordinary uniformity of the clockwork's operation both in the writing and the displaying apparatus" required was achieved by making use of his chronoscope's precision mechanics. Reproduced from Anonymous, "Der neue Buchstaben-Schreibtelegraph des Mechanikus Mathias Hipp aus Reutlingen in Württemberg," *Illustrirte Zeitung* 17: 491–92, here 492.

surprising, therefore, that in synchronizing the drums, Hipp went back to the construction he had already used in the chronoscope. Both drums of the telegraph were driven by clockworks with weights that were regulated, as they were in the chronoscope, by springs that performed one thousand oscillations per second.[63]

The *Illustrirte Zeitung* enthusiastically imagines the perspectives opened up by a writing telegraphy based on the chronoscope: "It will probably take only a few more years, and the earth will be covered with a network of lines for galvanic currents that allow for communications at the speed of thought"[64]—a prognosis that contained no trace of what had been emerging in Helmholtz's studies for about a year: namely the fact that nerve stimulations and even thoughts were by no means traveling as fast as electricity.

A short while later, Hipp was called on to join the newly founded state-run telegraphy workshop in Berne and soon became its sole director. The goal of the workshop was to ensure the assemblage, testing, and repair of all telegraphic apparatuses in Switzerland. Hipp's responsibilities included the appropriation of the material and especially the galvanic apparatuses needed for the national telegraph service.[65] Under his leadership, about thirty workers set out to open Switzerland to communication technology. After a series of conflicts concerning the way accounting was conducted at the Berne workshop (Hipp continued to produce and sell his own instruments), he ended up resigning in the summer of 1860. A short time later, he settled in Neuchâtel as an independent entrepreneur.

There, one of his first tasks consisted in equipping the newly built observatory with all electrical and telegraphic devices needed for a wide-ranging communication of time—especially for the telegraphic transmission of time signals to the clockmakers' workshops in the Jura mountains, which allowed for improving the quality even of simple clocks and watches.[66] Very quickly after the service had been established, it functioned so effectively that the Neuchâtel Post and Telegraph Office, too, procured its time signals from there. A short time later, these time signals were beginning to be relayed to all other telegraph stations in Switzerland. Starting from the Neuchâtel observatory, "for the first time, a standard or unified time for a larger geographical space" emerged in Switzerland.[67]

Hipp's fabrication and sale of electrical clock units were concrete contributions to this development, in Switzerland and beyond. In 1862, he delivered a first system of electrical clocks to the City of Geneva. Similar systems followed: in Neuchâtel, where it was used by the city and other "subscribers"; in Zurich in 1865, where 135 clocks were connected; in Winterthur; and in Königsberg (from which Helmholtz was again long gone) in 1869. Hipp later sold such clock systems to places as far away as Rome, London, and Philadelphia.[68]

Paris, however, was to decide in favor of a different system. Although Paris was a center of electromagnetic precision time measurement in the mid-1840s, the use of telegraphy for distributing precise time was slow to impose itself. Despite the establishment, in the mid-1850s, of a connection between the pendulum clock in the observatory and the central telegraph office, it was to take more than twenty years before the communication of time in Paris could be described as standardized. Not until the late 1870s did work on two more or less independent projects begin.

By 1880, the City of Paris had installed a system of electrical clocks. Public buildings like the town halls and schools of the city's districts were furnished with telegraphic time signals delivered by the observatory (Figure 24). The plan was to further extend this network of electrical clocks. In practice, however, the initiative was left to interested private citizens and institutions in the various neighborhoods and to the district town halls.[69] This self-organization of the electrical clock system was not particularly successful, especially since the private establishment and extension of another network for communicating time was working against it.

At the 1878 World's Fair in Paris, Austria had presented a system of pneumatic clocks that had been successfully operating in Vienna for some time. In the same year, interest grew in introducing a similar clock system in the French capital. At the end of 1878, the city approved a plan by the Société des Horlogers for such a system.[70] A short time later, an article in the *Revue chronométrique* discussed the technical details of its installation. Following the publication of an exhaustive description of Eugène Bourdon's hydro-pneumatic clocks and a somewhat shorter presentation of the pneumatic clock Eadweard Muybridge (later famous for his chronophotographs) had manufactured for San Francisco,[71] the Paris Society of Clockmakers finally decided to adopt the Vienna system.

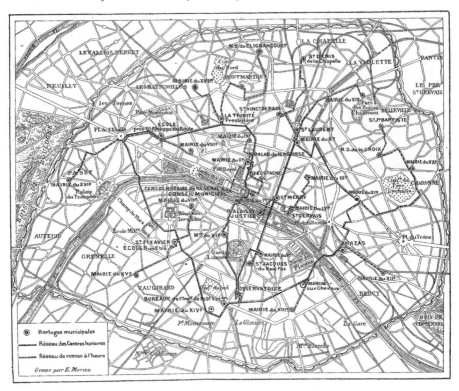

FIGURE 24. Map of the distribution of time to electrical clocks set up in public spaces in Paris (1881). The map shows two electric circuits: an Eastern one, connecting among others the city halls of the sixth and the twelfth *arrondissements*; a Western one, providing the city halls of the fifth, tenth, eleventh *arrondissements* and the *Hôtel de Ville* with time. The map shows a total of thirty locations for municipal clocks. Reproduced from Alfred Niaudet, "L'unification de l'heure à Paris," *La Nature* 9 (1881): 99–102, here 101.

In the early 1880s, the newly founded *Compagnie générale des horloges pneumatiques* began constructing such a system. Starting from a central clock that was readjusted daily via a telegraphic link to the observatory, time was distributed across the city partly through newly laid tubes, partly through already existing pressurized air tubes (the public pneumatic mail system had been in operation since 1879).

The first reports of the pneumatic clocks operating on the public squares and with a "good number" of subscribers date from March 1880 (Figure 25). There was satisfaction that the clocks, which were installed on tall columns, rendered time clearly visible in the streets of Paris: "The time stands out clearly from a sufficient distance."[72] The advantages of a time subscription in the comfort of one's home were emphasized as well: "Subscribers . . . receive, for a penny a day, the time of the Observatory at a minute's precision without ever having to worry about the rewinding, readjustment, or maintenance of their clocks."[73]

These clocks intervene in the work of Proust at a very precise point. In January 1910, about thirty years after the installation of the first pneumatic clocks in the city's public spaces (but still three years before the publication of the first part of the *Recherche*), Paris and the surrounding areas were hit by long and heavy rainfall. So much rain fell, the Seine burst its banks. Then, on the evening of January 22, all pneumatic clocks stopped at ten minutes past eleven o'clock: the basement in which the pressurized air was produced had been flooded.

Proust was evidently not a subscriber to pneumatically communicated time, since two days passed before he learned of the historical freeze of the Paris clocks. In a letter of January 25, Lionel Hauser told Proust about this momentous event—without, however, establishing a connection with Proust's work on his novel. Hauser was one of the author's financial advisers and merely wanted to prevent him from investing even more money in industrial shares. In this context, he pointed to the dangers to which industrial concerns like the Paris pressurized air company were exposed.[74]

In his answer to Hauser, Proust did not bring up the stopping of the pneumatic clocks, but he did express concern about whether the pneumatic mail system was running again. He often used this system to send his letters on their way across the city. His intensive correspondence guaranteed a variety of contacts with friends, relatives, and partners. And it concerned not only the sending of handwritten papers but also of photographs, especially the portraits Proust collected for use as raw materials for his writing.[75]

Additional contact with the outside world was established via a telephone that had been installed in Proust's apartment on the Boulevard Haussmann.

FIGURE 25. Pneumatic public clock in Paris, Place de la Madeleine, not far from Proust's *Boulevard Haussmann* apartment (1880). Reproduced from Edouard Hospitalier, "Les horloges pneumatiques: La distribution de l'heure à domicile," *La Nature* 8 (1880): 407–9, here 409.

It served, for example, to call cabs or to send the housekeeper's husband in a nearby café on errands.[76] Proust also gained access to the simultaneities of the metropolis via the daily papers. He wrote articles for *Le Figaro* and other Paris newspapers, whose business sections he, as stockowner, studied carefully.

Since 1913, moreover, Proust had been subscribing to the *théâtrophone*, a service offered in Paris since the 1890s. It allowed for assisting theater performances and concerts at home via the telephone (Figure 26). This device added an entire spectrum of acoustic material to the textual and visual raw materials Proust received in his apartment through the mail. A contemporary description explains the efficiency of the *théâtrophone*:

> As for the subscribers to the telephone network, they are in the best position since they, at home, without leaving their apartment or even their bed, may believe themselves transported to their favorite theater. When one knows the piece performed, one truly sees the setting and the actor's play again because one recognizes their voices and misses not a note, not a syllable.[77]

Not only visits of theater performances and concerts but even the everyday movements within Proust's sprawling apartment were mediated by a network of wires. The author's combined bedroom and study were connected with the other five rooms via electrical wires. Installed next to the table on

FIGURE 26. Main office of the *théâtrophone* in Paris (1892). Reproduced from Georges Mareschal, "Le théâtrophone," *La Nature* 20 (1892): 55–58, here 57.

which Proust kept his photographs was a switch cabinet that allowed him to turn the lights on and off, ring for the housekeeper, and operate an electric kettle—all this even while he remained in bed.

In the laboratories for physiology and psychology existing at the time, the situation was no doubt different. And yet these buildings contained a similar network of cables and wires to telephone between different rooms, to display lab time in the experimentation rooms, to present stimuli, and to measure (the time of) reactions in the soundproof chambers in which test subjects were placed. Pouillet's method, Hipp's chronoscope, and similar precision chronometers based on electromagnetism make the connection between psychophysiological research on time and the developing culture of telegraphy and telephony palpable. That culture has made Helmholtz and Proust contemporaries in the literal sense.

FIVE

Time to Publish

> The times are too restless, nothing is heeded.
>
> —ALEXANDER VON HUMBOLDT[1]

In practice, the research conducted by Helmholtz, Proust, and others who experimented with time in the nineteenth and early twentieth centuries referred to a collective labor on materializing simultaneity, an enterprise to which constructors of telegraphs and clockmakers, factory owners and newspaper companies, postal services and railroad companies contributed in their different ways and often independently of one another.[2] This is apparent in Pouillet's method but also in the drafts for the opening scene of Proust's *Recherche*.

These drafts show Marcel, lying in bed, reading the morning paper. Then, Proust has his narrator determine the time it takes his brain to identify a text he has written and now reads in *Le Figaro* as his own: "for a moment my thoughts, swept on by the impetus of this reaction, . . . continue to believe it isn't"—that is, to believe someone else has written the text.[3] In a next step, the issue is how, thanks to a newspaper, thanks to this product of a "mysterious process of multiplication," "thousands of wakened attentions" can emerge at the same time, in one morning: "At this moment each sentence that I extorted from myself flows not into my own mind, but into the minds of thousands on thousands of readers who have just woken up and opened the *Figaro.*"[4]

Inscribed in one of the primal scenes of the *Recherche*, then, we find astonishment about an author's "marvelous ideas" penetrating "at the same moment all the brains" engaged in reading a newspaper.[5] It is precisely this connection among spatially separate events operated by time that the novel, from the perspective of the narrator, ceaselessly questions in ever-new situations. And the definitive beginning of the novel makes clear that the questionability of these connections is to be measured in "seconds," "moments," and "blinks of an eye" rather than in days, weeks, years, or even decades.

The close connections Helmholtz's research machine entertained with the scientific-technological contexts of Berlin and Paris, too, approximated a simultaneification. At least they translated into a clear demand for temporal coordination in the winter of 1849–50. Given the apparently dynamic cooperation between Matteucci and Breguet, the process of publishing the semiotic output of Helmholtz's machine had to be sped up as much as possible. In his investigation, the Königsberg physiologist had found that a measurable time passes as stimulations propagate in a nerve. Now it was up to him to not waste any time in publishing this finding in order to have it penetrate "all the brains" busy with reading scientific journals—even if Helmholtz, unlike Proust writing his draft, could be certain that this would not, strictly speaking, take place all "at the same moment." Not weighed down by book projects, he therefore authored his famous "Preliminary Report."

In principle, drafting such a report was nothing unusual. Since the 1830s, the audience of German language physiology journals had been accustomed to the genre of "preliminary" messages, remarks or, precisely, "reports," and well-known scientists such as Christian Gottfried Ehrenberg, Jan Evangelista Purkyně, Robert Remak, and Johannes Müller had made use of the genre. Helmholtz, however, was not merely concerned with providing access to the first results of newly begun and intermittently interrupted research to be confirmed or rejected again by later studies—his own or those of other physiologists. His report also aimed at making an individual "mark" in his science, at leaving a personal date in the history of physiology: not, like du Bois-Reymond, with a comprehensive whole, a book or textbook, but, more like Matteucci, with a fast publication, a journal article, or a preliminary, and therefore also precursory, report.

This ambition is clearly visible in the very first sentence of the letter to du Bois-Reymond that accompanied the manuscript. Helmholtz asked his friend in Berlin to "officially present it [the "Preliminary Report"] in the Physical Society and to deposit it in its files to preserve priority."[6] The priority to be thus established did not abstractly refer to Müller, whose assessment in the *Elements of Physiology* of the outlook for precision time measurement had turned out, in light of recent technological developments, to be rather too somber. For that, Helmholtz's respect for his teacher was too great, but Müller's current research was also too unrelated both in form and content. Neither, however, did Helmholtz's claim to priority concern the concrete project of the physiological measurement of time proposed by du Bois-Reymond in 1845 but never put into practice. Instead, the phrase "preserve priority" positions and temporalizes Helmholtz's work with a precise view to ongoing research activity on an international scale, especially as regards the collaboration of Matteucci and Breguet.

In the mid-1840s, du Bois-Reymond had been forced to watch helplessly while Matteucci, in the Paris Academy's *Comptes rendus*, claimed for himself the establishment of electrophysiological facts that he, du Bois-Reymond, believed he had described already two years earlier. After the first two volumes of his *Studies* had been published, du Bois-Reymond could finally react. In the spring of 1850, he traveled to Paris to enter into a public confrontation with Matteucci, both in the forum of the Academy's Monday meetings and in the printed forum of the *Comptes rendus*. As we shall see, Helmholtz's "Preliminary Report" entered these interventions as an additional element in the demonstration of the priority, and thus also the predominance, of the physiological research undertaken by students of Müller—provided communication and publication took place quickly, and in French, too.

In addition to the letter to du Bois-Reymond accompanying the "Preliminary Report" in the middle of January 1850, Helmholtz sent similar letters to Johannes Müller and Alexander von Humboldt, hoping to have the "Preliminary Report" forwarded to the Academies in Berlin and Paris, respectively. The effect of these dispatches from Königsberg was a veritable cascade of communications in the German-speaking world. Müller opened the series, reading Helmholtz's "Preliminary Report" before the Academy of Sciences in Berlin on January 21. Ten days later, it was du Bois-Reymond's turn. He read Helmholtz's report before the Physical Society on February 1. Then followed documentation in print. In the course of the year, the "Preliminary Report" was published in a total of three German-language journals, first in the *Bericht über die zur Bekanntmachung geeigneten Verhandlungen der Königlich Preussischen Akademie der Wissenschaften zu Berlin*; then in the *Archiv für Anatomie, Physiologie und wissenschaftliche Medicin*, edited by Müller; and finally in Poggendorff's *Annalen der Physik und Chemie*.[7]

The reinvention of an experiment

The result of this orchestrated communications operation, quickly and widely spread as these publications were, was harrowing. In March of 1850, a clearly sobered-up du Bois-Reymond wrote, "with pride and sorrow," to Helmholtz that the studies described in the report had been "understood and appreciated" only by himself.[8] Yet according to du Bois-Reymond, the reason did not lie in the radical newness of what was reported but in—the "Preliminary Report." He writes to Helmholtz:

> It's because—don't take this the wrong way—you presented the matter in so exceedingly obscure a manner that your report could at most count for a short instruction for reinventing the method. The result was that Müller did not reinvent it and that after his presentation, the Academy members imagined you failed to eliminate the time passing in the process that takes place in the muscle. I had to enlighten them one after the other, Riess, Dove, Magnus, Poggendorff, Mitscherlich, finally Müller himself, who really didn't want to take it on at all.[9]

"Reinventing the method" is no doubt a remarkable formulation. What is usually understood to be a creative achievement, often even as the immediate production of a new insight or a new technique, here appears very much as a reproducible phenomenon. A "short instruction" thus not only makes it possible in principle to make an invention once again. Du Bois-Reymond is also and concretely certain that the result of the innovation reproduced corresponds to the initial innovation—which may count as further evidence for what du Bois-Reymond called the "interlocking manual labor" of the two scientists in Berlin and Königsberg.

To a certain extent, it may. For although Helmholtz was certainly aware of the difficulties of the "Preliminary Report," what was needed in his eyes was not a complete repetition but merely the supplementation of a necessarily compressed presentation. Thus he responded to du Bois-Reymond's criticism: "The redaction of such a note is so difficult because its main purpose is to serve as a short hint for the expert who has to add details to it by invention."[10] Nonetheless, whether adding by invention (that is, by an independent supplementation of details) or by reinvention, on the level of communication the result was the same. In Berlin, the "Preliminary Report" was received only with reservations.

Ironically, this confirmed, implicitly and on a different level, as well as on a larger scale, just what Helmholtz had "found" at the very beginning of his studies on the temporal course of muscle activity. The communication of a scientific fact also does not take effect at the moment of stimulation, of communication, but occurs, for the most part, after the stimulation has ended. Just as in the case of physiological communications in the body, public communication among physiologists had its "quiet dark times." Or, in the analogy Marey uses in a later work to explain the *temps perdu*: A letter may very well have reached its destination. But that does not mean that it has been received.[11]

A similar delay was taking shape in the French context. The details of the "Preliminary Report" remained obscure to Alexander von Humboldt, who was to transmit the report to the Paris Academy. This must have been particularly vexing. For, on the one hand, the Académie des sciences at the time had the reputation of being, as Siemens retrospectively remarks, "the world's premier scientific agency."[12] On the other hand, and most important, this agency had since its foundation considered establishing "the right dating of discoveries" to be one of its central tasks.[13] But there was nothing doing. Von Humboldt stalled. He was puzzled. He hesitated. Bewildered, almost with disbelief, he wrote to du Bois-Reymond on January 18 that the method of measuring time that had been applied by Helmholtz was not explained anywhere in the "Preliminary Report":

> I'll gladly send the excellent Helmholtz's attachment [i.e., the "Preliminary Report"] to Paris, but I find myself confronted with this form completely incomprehensible to me: one speaks of 14/10000 of a second without indicating how such a portion of time is being measured.[14]

Once more, du Bois-Reymond tries to act as enlightener. Against the background of his "reinvention" of Helmholtz's method and the prospect of soon entering publicly into the polemic with Matteucci, he offered his services as translator. He was willing to produce a French version of the text written in German that could be published in the *Comptes rendus*. At the same time, he suggested "reworking" Helmholtz's text "for comprehensibility."[15]

And indeed, the German and the French versions of Helmholtz's report differ significantly. First, in terms of their arrangement, Helmholtz organizes his report into six paragraphs, whereas du Bois-Reymond turns them into ten and assigns numbers to two of them. Helmholtz packages his message in nineteen sentences, du Bois-Reymond divides them into thirty-five. Helmholtz places the summary of his numerical results at the beginning, du Bois-Reymond moves that paragraph to the end of the text.

There are also differences in content. In his description of the experimental setup, Helmholtz focuses on processes ("The stimulation of the nerve took place . . . ," "The current meanwhile circulated in the multiplicator . . . ," "The deflection the current imparts to the multiplicator's magnet bar in its transition . . . ," etc.);[16] du Bois-Reymond, by contrast, foregrounds the spatial distribution and arrangement of the experiments' components ("I insert the sciatic plexus into the electric circuit . . . ," "The muscle is arranged in such a way that . . . ," etc.).[17] Helmholtz embeds the discussion of possible measurement errors from the outset in his presentation of each step of the experiment; du Bois-Reymond collects them at the end of his description of the experimental setup.

Above all, however, du Bois-Reymond clearly brings out the variation and subtraction method that had made the transition from measuring muscle contractions to measuring nerve stimulations possible. Helmholtz, in this respect, had left it as something of an aside ("What emerges from the magnet's deflections is that the same mechanical effect sets in somewhat earlier when the lower end of the nerve is stimulated as when the upper end is stimulated"),[18] which may have been one of the reasons Müller and others had at first believed he "had failed to eliminate" the contraction time. Du Bois-Reymond, in contrast, distributed repeated explanations across the conclusion of the translated version of the "Preliminary Report," meant to clarify the procedure of varying and subtracting that was decisive for the nerve time measurements ("By varying the length of the stimulus path in the nerve . . . ," "If one alternately stimulates the upper and the lower parts of the nerve . . . ," "It is therefore sufficient to compare the numbers of the two experiments . . . ," and so on).[19]

At several points, du Bois-Reymond also elaborates on Helmholtz's description, especially by means of other, earlier authors who—with the exception of Pouillet—go entirely unmentioned in Helmholtz's version. In a way, du Bois-Reymond thus historicizes Helmholtz's short text even if the name of Matteucci—as would be expected—does not come up. Where Helmholtz, for example, writes: "I measured the deflection with mirror and telescope,"[20] we read in du Bois-Reymond's version: "The measurements were taken, as in the apparatuses of Gauss and Weber, with the help of a mirror and a telescope."[21] Helmholtz remarks in a subordinate clause: "which corresponds to the familiar experiences made with cut-out nerves dying off starting at the central end";[22] du Bois-Reymond reformulates: "This phenomenon corresponds to what Valli and Ritter have observed in frogs prepared in Galvani's manner."[23]

A footnote that converts the 1/100 and 1/1000 seconds Helmholtz had measured into meters per second is also added. *This* "translation" was likely motivated not simply by the wish to present the results as a whole more vividly but also to make it easier to compare them with the measurements of the velocity of light and electricity von Humboldt had been interested in in the recently published third volume of his *Kosmos*. There, the predominant unit was "meters per second."[24]

Finally, he was satisfied. For von Humboldt, the French version of Helmholtz's report was a "new copy" (*neue Abschrift*).[25] Like the term "reinvention," this formulation is remarkable, as if the French version of the "Preliminary Report" were a copy and, at the same time, an original. In a way, the copy indeed *was* an original, since it was based on du Bois-Reymond in Berlin reinventing the experiments Helmholtz had performed in Königsberg. Ultimately, no one could know how many differences this repetition had introduced.

Von Humboldt, of course, meant something different. Writing to Helmholtz, he explained that translating the report had "rectified some expressions and clarified them."[26] Du Bois-Reymond similarly told his friend and colleague that in his translation and revision he had "not added a single detail" but "strictly" kept to the "given"—which he had, "however, developed inductively."[27] This was supposed to mean that he had presented the experimental procedure in the order in which it had presumably proceeded in reality—although du Bois-Reymond did presume that this order was also logically consistent.[28]

On the level of the text, this "inductive" moment of reconstruction obviously refers to moving the first paragraph from the beginning to the end of Helmholtz's "Preliminary Report." In the French version of this text, the experiment's results appear in the form of numbers only after the description of the experimental setup and of the way the measurement experiments were conducted. Beyond that, however, the inductive moment also concerns the presentation of the setup as a whole. Du Bois-Reymond had structured it more clearly into successive steps and at the same time toned down the pervading sense of processes and problems. The translator left it to his friend in Königsberg to examine and evaluate all other details of his two-fold mediation—with one exception: "The remark on the velocity per second is not mine but Humboldt's."[29]

In this form—rewritten by du Bois-Reymond to be logically and historically consistent and with an explanatory footnote by von Humboldt—the "Preliminary Report" was presented during the session of the Académie des sciences that took place on February 25, 1850, as "Note sur la vitesse de propagation de l'agent nerveux dans les nerfs rachidiennes," authored by Helmholtz. Incidentally, this happened immediately after the presentation of a communication from the apparently omnipresent Matteucci. On this occasion, however, he discussed a topic in physics, namely the voltaic arc.[30]

Publicly in limbo

All the modifications and additions to Helmholtz's "Preliminary Report" could do nothing to change the fact that in the French context, too, it had no immediate effect. Picking up on Marey's analogy again, we might say the communication had reached its destination but was not received. Or rather, it was received, but not at all in the way one would have hoped. The Paris press immediately mocked Helmholtz's time measurements.

Only two days after the presentation of the "Note" in the Académie des sciences, the liberal daily newspaper *Le National* ran a short report on these measurements as part of its monthly science review. After a rather sober summary of the results contained in the "Note," the authors, writing under the byline "D. et T.," remark: "No need to add that it was the frogs who had to pay the price of this work. Frog aside, Monsieur Helmottz's [*sic*] work leaves something to be desired."[31]

This apodictic formulation was obviously directed against the human agent who had presented his research to the Academy. Yet what exactly was meant by "something to be desired" remains largely obscure to us readers today. What we can say, however, is this: The report written up by D. and T. was embedded in a reflection on the Académie des sciences. This reflection was critical of science, praising the institution's "beautiful name" but questioning the "matter."

Alluding to La Fontaine's fable "Simonides Saved by the Gods," the authors compare the Académie with a hero about whom there is nothing to say and the reporter with a poet who is quite willing to sing the praise of an athlete but, confronted with "a matter not overly fertile"[32] is quickly forced to digress. Then, D. and T. add: "Our own hero"—the Academy—"does nothing, says nothing, hears nothing—he, who knows so many things!"[33] Month after month, therefore, one is forced to publish "a summary of little news"—a complaint that overlaps with du Bois-Reymond's earlier views on scientific journals in surprising ways. In fact, the criticism of the journals, overwhelmed with disconnected facts and observations, that the electrophysiologist had voiced a few years earlier is here turned against Helmholtz's "Note," which had been prepared and positioned so carefully. In any event, it was this "Note" that figured first among the scientific news in

Le National's science section. What, for D. and T., was missing in the note, was, quite obviously, something that would go beyond the monotony of everyday scientific news and thereby contribute to saving the "helpless Academy."[34]

Helmholtz was not impressed. He could not read the article by "D. et T." because *Le National* was not available in far-off East Prussia.[35] But du Bois-Reymond had told him that he had been "made fun of" in that newspaper.[36] This sufficed to elicit a dispassionate reaction that situated the whole affair on the level of national and cultural difference. To his father, Helmholtz wrote: "A benevolent reception by the French of such things from Germans just isn't possible, and the preliminary purpose of drawing their attention has been fulfilled."[37] The "*preliminary* purpose" of the "*Preliminary* Report": This doubly provisional character sums up the dating, the temporalization, that Helmholtz was most concerned with in his first communication to the Paris Academy.

Du Bois-Reymond reacted less coolly. To him, the whole affair seemed to repeat a familiar pattern. One year earlier, in May 1849, the first reports about his own electrophysiological experiments had appeared in the *Comptes rendus*. Von Humboldt had informally yet repeatedly informed the Académie about the progress of the young Berlin electrophysiologist's work. In the spring of 1849, he had been particularly impressed with a demonstrative experiment of du Bois-Reymond's, in which du Bois-Reymond had deflected a galvanometer needle by contracting his arm muscles. Von Humboldt had written to his friend Arago about this several times. A note on this experiment that von Humboldt and du Bois-Reymond published together in the *Comptes rendus* on May 21, 1849, however, met with mixed responses.

The science section of *Le National* considered the experiment to be a "great discovery" on the part of the "physicist from Berlin."[38] The *Journal des débats*, however, ran a scathing criticism that was to be taken all the more seriously for having been written by a member of the Academy, Léon Foucault. Foucault presents the electrophysiologist from Berlin as a "beginner eager for discoveries" put on entirely the wrong track by an oversensitive galvanometer. Du Bois-Reymond, he writes, seriously believed the human will to be capable of moving a magnetic needle: "The very proposition . . . aims at the marvelous and bears the mark of the German mind that formulated it."[39] Subsequently, another journal, the *Gazette des hôpitaux civils et militaires*, polemicized against von Humboldt in particular, demanding that in his descriptions of du Bois-Reymond's experiment he "return to rigorous language" and that he avoid abetting "charlatanism."[40]

Given the curt newspaper report on Helmholtz's time measurements, du Bois-Reymond now seemed to fear that a similar polemic would develop against the Königsberg physiologist's research. Although the report about Helmholtz published by D. and T. in *Le National* was much less pointed than the earlier criticism by Foucault, in Paris there seemed to be another attack about to be mounted on yet another scientist from the Müller school. This, at least, was the assessment of the situation du Bois-Reymond forwarded from Berlin to Königsberg. The Paris press seemed to sound off against Helmholtz "just as they did against me back then."[41] The only concrete indication of such treatment, however, was that *Le National* had turned Helmholtz's name into "Helmottz" and introduced him as "Professor of Pathology," not as physiologist. "J'y mettrais ordre"—I'll settle this—du Bois-Reymond confidently and eloquently promised, although he had not really been asked to do so. On March 15, 1850, he left Berlin for the "Capital of the Nineteenth Century."

The first and the main reason for this trip was to present the first two volumes of his *Studies on Animal Electricity*, which had been out for about a year.[42] Backed by this "whole," du Bois-Reymond now finally wanted to strike against Matteucci. He wanted not only to give expression to his long-retained claims to priority but also to justify them in detail. The second reason for the journey was the misunderstandings and rumors that had been circulating in Paris since the summer of 1849 about the demonstrative experiment on the deflection of the galvanometer needle by the arm muscle. Like the presentation of the *Studies*, this, too, was a scientific mission and, at the same time, a matter of polemics. In a letter to Helmholtz, du Bois-Reymond accordingly spoke of wanting to "tear the rabble to pieces" in the City of Lights, which Helmholtz in turn translates as "enlightening the Parisians heads as to German science."[43] The third, supplementary reason for the electrophysiologist's Paris trip was to explain Helmholtz's measurements concerning the propagation speed of nerve stimulations—a matter close to his heart, probably also because he had played a major part in drafting the relevant announcement in the *Comptes rendus.*

Shortly after du Bois-Reymond arrived in Paris, it turned out that the trip had a fourth, unforeseeable purpose. On March 28, Werner Siemens arrived in the city. Siemens was planning to present to the Académie the model of an electrical telegraph he had developed. Right after his arrival, he had sent his card to du Bois-Reymond, moved into the same hotel near the Jardin des Plantes shortly afterward, and spent a lot of time in the following four weeks with the physiologist, whom he knew well from the Physical Society in Berlin. As du Bois-Reymond's travel diary documents, the two had dinner together, visited the collection of machines in the *Conservatoire des Arts et Métiers*, and even assisted a *séance magnétique* with the then-famous medium Prudence Bernard in the Bonne Nouvelle Theater.[44]

As he did for Helmholtz, du Bois-Reymond also acted as enlightener on behalf of Siemens. In the roughly two months he spent in Paris, he translated the comprehensive exposition of Siemens's telegraph system into French. A short summary of this "Mémoire" was presented in the Academy during its session of April 15 and subsequently published in the *Comptes rendus.* The full text, which ran to almost fifty manuscript pages, was later published in the *Annales de Chimie et de Physique.* This was the basis of a book publication of Siemens's "Mémoire" that followed a short while later in Berlin. Once back from Paris, du Bois-Reymond was to send a copy of this work—"written by me in Paris" in French—to Helmholtz.[45]

From an academic point of view, this support of Siemens was quite successful. As early as the end of April, the commission set up by the Académie declared the superiority of the electrical telegraph constructed by Siemens over other models presented and discussed in Paris at the time.

Du Bois-Reymond's own business, too, made progress, albeit more slowly. On March 25 and April 8, he lectured before the Académie, presenting the major results of the two volumes of the *Studies on Animal Electricity* and recalling his "Preliminary Summary" of 1843 (without, however, mentioning Matteucci). It didn't take long for the Pisa physicist to understand and to react. On April 22, Matteucci claimed "priority" for the contributions to electrophysiology described by du Bois-Reymond. Du Bois-Reymond countered with two "Answers" he read in the sessions of April 29 and May 6.

Matteucci objected again, on June 3. At this point, however, du Bois-Reymond had already left Paris. He therefore was not present when, on July 15, a commission presented a report that took stock of the results du Bois-Reymond had presented and contained a statement on the polemic with Matteucci. But back in Berlin, du Bois-Reymond counted the detailed report in the *Comptes rendus* as a victory: "Matteucci and his objections are rejected . . . my laws and terminology officially accepted," he writes to Helmholtz.[46] Then he pursued the matter one last time, in a concluding answer to Matteucci in the *Comptes rendus* of July 22.

As for Helmholtz's time experiments, du Bois-Reymond achieved no such success, although he had seized the very first opportunity for speaking up in this matter. In his first lecture at the Académie, on March 25, he had concentrated on presenting the historical and methodological preconditions of electrophysiological work, in a way similar to his introduction to the first volume of his *Studies*. After some remarks on eighteenth-century stimulation theory and on the objections raised already by von Haller and Lecat against equating animal spirits with electricity, du Bois-Reymond jumped to the present time, to the work done by his friend and colleague in Königsberg:

> Mr. Helmholtz's experiments have shown that the propagation speed of the nervous agent is more than ten times less than that of sound in the air and thus more than one hundred times less than that of electricity in copper! . . . Accordingly, it is impossible to identify the nervous agent with electricity.[47]

Yet for all his enthusiasm, he achieved little with his lecture. In fact, the members of the Academy in attendance were somewhat surprised by du Bois-Reymond's long-winded historical discussions—unpleasantly surprised. For the publication in the *Comptes rendus*, du Bois-Reymond therefore reduced the historical introduction to a minimum and concentrated on presenting the effect contractions have on the muscle current, one of the core issues taken up in the *Studies*.

As for Helmholtz's time experiments, which he had mentioned in his introduction, it seems that the reactions were even more devastating. He was to report later to Helmholtz, "Your measuring of the velocity of the nervous agent was ridiculed in the Academy."[48] Still in Paris and rather stunned, he wrote to Carl Ludwig about his fellow physiologists in the city: "They think Helmholtz is a madman."[49] Nonetheless, he had another go at it in a lecture he gave in the Société Philomatique on March 30. Yet there, too, the response seems to have been muted: "Although nobody dared to object publicly, I was plagued silently by the most stupid of doubts and objections."[50]

All in all, these reactions are somewhat surprising. For Paris—at least the Paris that du Bois-Reymond visited—had become a center of experimentation on precision time measurement since the chronoscope debates between Pouillet, Breguet, and Wheatstone (see chapter 4). During his two-months stay in Paris, du Bois-Reymond attended a whole series of pertinent lectures: for example at the Académie on April 15, where Hippolyte Fizeau spoke on the velocity of the electrical current; at the Sorbonne on April 30, where Pouillet talked about the speed of light; and, once again at the Académie on May 6, where Foucault gave a lecture on the speed of light in air and in water. Arago showed him a rotating mirror turning at 1,600 revolutions a second. Wheatstone had used a similar mirror in his measurements of the velocity of electricity, a technique that Fizeau and Foucault adopted as well.[51]

As familiar as it now was to take such high-speed measurements and as uncontested their scientific interest now seems to have been, so unusual, evidently, was their extension from physics to physiology. Helmholtz's experiments in this vein were seen as the experiments of a "madman"—quite a strong statement. As we have seen (chapter 1), Helmholtz reacted with a second communication that was read in the Paris Academy in September 1851 and printed soon afterward in the *Comptes rendus* as the "Second Note." As in the case of du Bois-Reymond and Siemens, a commission was formed to further look into and assess the content of Helmholtz's message. Yet this commission never published a report.

Instead, a short summary of the "Measurements" (the ninety-page treatise from 1850) was published in French in 1855—but not in a physiology periodical. It was published in the *Annales de Chimie et de Physique*, in which Siemens's description of the electrical telegraph had appeared as well. That same year, Jules Béclard's physiology manual mentions Helmholtz's time-measurement experiments.[52] The reception of these experiments did not gain traction in the French-speaking world until the mid-1860s—thanks to Marey's publications. Employing once more the metaphor he himself coined, it was Marey who took on the task of delivering the scientific mail that had arrived in Paris quite some time before. Yet—to continue the metaphor—in the package he delivered, the graphical method was sitting on top. Pouillet's method was hardly mentioned anymore.

SIX

Messages from the Big Toe

What we do we do not know in an immediate manner.

—HERMANN HELMHOLTZ[1]

While, in the spring of 1850, du Bois-Reymond was in Paris, promoting Helmholtz's time experiments, the research machine in Königsberg was fully engaged in drifting. A short time after the "Preliminary Report" had been drafted and sent off, the assemblage of frog frame, galvano-chronometer, and reading telescope was already moving from working with weakened winter frogs to operating with fresh spring frogs, and had passed from dealing with nerve stimulation as a temporally conditioned phenomenon to engaging with this stimulation as a phenomenon that was, equally, thermally conditioned.

Helmholtz reported on these shifts in great detail in the "Measurements," the ninety-page treatise he finished writing in June 1850. What he does not mention in the "Measurements" was a third drift running parallel with the first two, a drift that would turn out to be decisive for the history of psychophysiological research on time. This drift led to an event that would recall itself "forwardly" (*vorlings*) in later research machines and in different places. Yet this event was neither a further achievement for which priority could be claimed nor did it provide a convincing answer to a precisely asked question. Rather, it constituted an exemplary opening that provided drafts for solving problems that had yet to be raised in concrete ways.

For, in parallel with the shifts just mentioned, the biological components of Helmholtz's research machine changed in an even more fundamental way: Some weeks after drafting the "Preliminary Report," Helmholtz began to extend his time-measuring experiments on frogs to human beings. At the same time, he added sensory nerves to the motor nerves as elements of his research machine. Already at the beginning of April 1850, while du Bois-Reymond was still in Paris with Siemens, Helmholtz reports to Berlin: "I have taken time measurements on myself and other human beings, which to me seem to establish the propagation speed in the motor and sensory fibers of the human being as one of 50 to 60 m[eters]."[2]

Exit the frog, enter the human: this was a reality as early as spring 1850, a reality in which Olga Helmholtz once again actively participated. This time, she functioned not only as assistant and recorder but in a new double capacity as experimenter and test subject. As Helmholtz writes retrospectively, she had become a laboratory worker who was "now so adept in physiology that she was able to conduct series of experiments on the velocity of stimulation in the nerves on herself."[3]

Yet, just like the earlier transition from chronography to chronoscopy, the ejection of the frog from the research machine and the human being's entry into it was more than a mere replacement, more than a question of method or technique. It constituted a supplementation and extension of an experimental practice that led to a displacement of the entire research enterprise pursued until then. In distinction from the frog time measurements, Helmholtz in his human subject experiments did not work with hanging muscles and exposed nerves separated from the rest of the organism. He included the whole human being, both in applying the contacts for the electric stimulus and in converting the stimulus into contractions. What effect, exactly, this preservation of their embeddedness in the surrounding body had on how muscles and nerves functioned could not be conclusively assessed. Yet despite the significant modification of the experimental setup, Helmholtz stuck closely to the procedures of his earlier frog experiments.

He first turned his attention to the human being's motor nerves. His experimental setup was modified such that "small electric shocks" could be sent "through various points of the median nerve on the inner side of the biceps."[4] Clearly impressed, Helmholtz describes to du Bois-Reymond how this gave rise to "the most beautiful involuntary twitching in the forearm flexors."[5]

The hand movements that resulted from these contractions in muscles of the forearm could then be used to interrupt the time-measuring current—in analogy to the frog's calf muscle whose contractions had opened a contact in the earlier version of Helmholtz's machine. If now, according to the familiar pattern, different points of the median nerve were stimulated and the corresponding spatial and temporal differences were recorded, the propagation speed in human motor nerves could be derived from these variations and subsequent subtractions. So far, so good.

In an additional series of experiments, however, Helmholtz moved away from the model of the frog sample. He turned away from human motor nerves and toward human sensory nerves. At first, this seems just a change of direction. Instead of tracing the stimulation from the inside to the outside, from the nerve to the muscle, it was now observed in the opposite direction, from the outside in. In fact, however, this constituted a fundamental change in strategy, for the opposite direction led from stimulating a point on the skin—for example, on the neck, hand, or foot—to the mouth or a moving hand; it led from outside via inside back to outside. Put differently, in the second series of experiments, the entire human body, including brain and spinal chord, became a component of a time-measuring experimental setup that, as Helmholtz writes, "must actually make use of" this body to obtain the desired time measurements.[6]

On the level of scientific practice, the opening of Helmholtz's experimental setup that comes with this reversal constitutes the decisive event of psychophysiological time research in the nineteenth century. This opening detached the variation and subtraction method from the narrow confines of a *thing*, the prepared frog sample, and extended it to the *space* of the entire human body. Contrary to Johannes Müller's expectations, the surfaces of this body were now to grant time-measuring physiologists access to the brain and the nervous system on the one hand and to the functions of consciousness on the other. Helmholtz thus passed from physiology in the restricted sense to experimental psychology long before there were laboratories for that purpose.

In the relevant experiments, the electric current acted as a "stimulating shock" on a particular point on the skin while the test subject, "with fixed attention," was "ready, as soon as [she] felt the shock, to interrupt the time-measuring current with [her] hand or, even better, with [her] teeth."[7] Helmholtz then stimulated another point on the skin while the test subject, "with fixed attention," was again to interrupt the time-measuring current. He was then able to derive the propagation speed of the stimulation in the sensory nerves from the differences in times and from the various distances of the points on the skin from one another.

The difference, compared to the first series of experiments on the motor nerves, was, first, that the contact point for stimulations was no longer just the arm but, in principle, the entire body. Instead of merely varying the way the electrodes were placed on the inside of the arm, it was now possible to work with the differences between foot and chest or between finger and neck. The advantage was clear. It was possible to establish greater distances between the nerve points and thus achieve "better success" in the measurements.[8]

Second, however, the experiments were now no longer working with involuntary contractions but with conscious reactions. For taking precise measurements, this could be a disadvantage. Even more than in his experiments with prepared frog samples, Helmholtz had to anticipate disturbances in the course of his time measurements caused by a number of factors, including the test subjects' individual "freshness," the temperature in the laboratory, and variations in the fixedness of individuals' attention. Nonetheless, he had obtained promising results as early as April 1850: "All of the times measured vary on different days and in different people; the differences of time for toes and sacrum or for finger and neck etc., however, are constant such that they can, with great probability, be related to the nerve distances."[9] Thus it was in the end the constancy of the measurement results, which surprised the experimenter, that allowed for access to the object being measured. This object, however, was by no means identical with the sensory nerves: It also included the functions of the brain.

Why this renewed drifting? Why did Helmholtz, in the spring of 1850, extend his experimental time measurements from frogs to human beings? We can name at least three reasons. First, reasons of a technical kind: Given the novel time-measuring procedure he had developed, Helmholtz was concerned with applying it "to as many cases as possible"[10] to demonstrate the procedure's productivity and to further explore its robustness. Second, of an epistemological kind: Helmholtz sought to move from "the simplest cases" in frog physiology to the more complex situation in human physiology,[11] not least in order to emphasize the validity and therefore the relevance of his earlier animal experiments. Third, of an aesthetic kind: As Helmholtz had learned thanks to the reticent reception of his "Preliminary Report" in Berlin and Paris, the results of electrophysiological laboratory research remained difficult to communicate even to the qualified academic public.

Du Bois-Reymond had reacted to these problems early on by transferring his 1843 "discovery," the so-called negative variation of muscle currents in frogs, onto the terrain of human physiology in order to illustrate it. That was the background for instituting and performing the quasi-gymnastic experiment of effecting a needle deflection on the galvanometer by contracting the arm muscles. At the end of the 1840s, Helmholtz, too, had attended such demonstrations, and he may well have had these in mind when he began his own human subject research in Königsberg.

The continued work on these experiments also ran parallel, at least in part, with the written elaboration of his "experiments on human beings"[12] that du Bois-Reymond was working on with a view of publishing it as part of the *Studies on Animal Electricity*. As late as February 1852, du Bois-Reymond writes to Helmholtz: "You'll be surprised when I tell you that I'm still struggling with the human currents."[13] Like an echo, Helmholtz answers: "My measurements of human time give much reason for hope."[14]

In public, Helmholtz in the end only mentioned his time experiments on human beings in a popular lecture. This, too, speaks to the aesthetic motivation of these experiments. The lecture he gave on December 13, 1850, before the Königsberg Physical-Economical Society, began with a presentation of the new "methods of measuring very small portions of time," which also recapitulated the priority debate that had concerned the question of electromagnetic precision time measurement from Pouillet via Breguet and Wheatstone all the way to Siemens. Helmholtz then turned to the studies of the motor nerves of frogs he had presented in detail in the "Measurements" to ask in the conclusion: "How stands the question in the case of man?"[15]

The answer he gives concentrates on the series of experiments on human sensory nerves: "The message [*Nachricht*] of an impression made upon the ends of the nerves in communication with the skin is transmitted to the brain with a velocity which does not vary in different individuals, nor at different times, of about 60 metres (195 feet) per second."[16] This was the discursive correlate of performing time-measuring experiments that "must actually make use of" the human body. Then Helmholtz added a second result he had not mentioned in his letters to du Bois-Reymond: "Arrived at the brain, an interval of about one-tenth of a second passes before the will, even when the attention is strung to the uttermost, is able to give the command to the nerves that certain muscles shall execute a certain motion."[17] This was a new interval that illustrated the expanded possibilities opened by his time experiments. Besides the activity of muscles and nerves, it was now also possible to capture brain activity in temporal terms. In other words, Helmholtz paved the way for characterizing additional psychophysiological structures and functions by means of their intermittent temporality.

The telegraph analogy

As his use of the term "messages" indicates below, Helmholtz embedded the report on his psychophysiological time measurements in an analogy with telegraphy. Even before he begins describing his research in the lecture before the Physical-Economical Society, he declares that

> the nervous fibres might be compared with the wires of the electric telegraph, which in an instant transmit all messages [*jede Nachricht*] from the extremities of the land to the governing centre, and then in like manner communicate the will of the ruling power to every distinct portion of the land.[18]

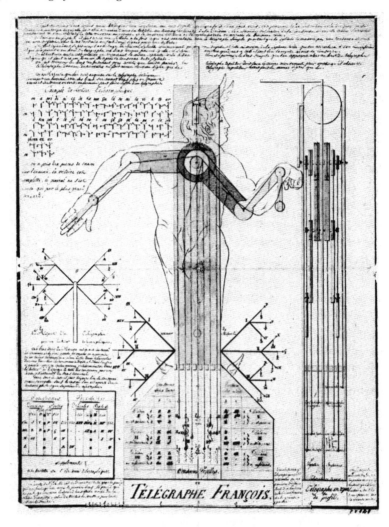

FIGURE 27. French depiction of telegraph code trees (ca. 1800). Reprinted with permission from Deutsches Museum, Bildstelle.

Around 1850, it was by no means unusual to draw on this analogy. In the early nineteenth century, the wooden signal stations of optic telegraphy had often been compared with the form of the human body. In some French depictions of semaphores, these were literally written on the body of a Hermes figure (Figure 27). Conversely, a drawing by Goya, *Telégrafo* (ca. 1825–28), shows a signalizing acrobat standing on his head, his legs serving to emit signals (Figure 28). Here, the entire human body turns into a telegraph station.

A palpable connection between communications technology and brain and nerve science, however, had been established as far back as 1809, when the anatomist and anthropologist Samuel Thomas von Soemmerring presented his electric letter telegraph to the public. In the writings of Carl Gustav Carus, Gustav Theodor Fechner, Johannes Müller, and other early nineteenth-century philosophers, physicists, and physiologists, this connection is translated into a conceptual analogy of (telegraph) wire and nerve.[19] Thanks to the close connection between telegraphy and the electromagnetism-based time measurement in Wheatstone, Breguet, and Siemens, the analogy became concrete once more in Helmholtz.

In other words, when Helmholtz talks about the telegraph, he does not only make a comparison, he allows a technology whose basics he was familiar with from experience and from employing it in his lab to intervene in his text. He is therefore fully aware that the comparison doesn't quite work—and this awareness, precisely, constitutes his aesthetic, his didactic calculation. What his time experiments on human beings so impressively demonstrated was the fact, precisely, that the propagation of nerve stimulations does *not* take place "instantaneously" but significantly more slowly, with delays and the interpolation of "interims."[20] While the material culture of electromagnetism and telegraphy constituted a positive technological precondition for precisely measuring time in the physiological laboratory, in the lecture hall, it constituted the negative foil against which the halting "transmission"[21] of messages within the human body could be vividly described.

The analogy with telegraphy served yet another purpose. Helmholtz also employed it to introduce a certain kind of structure. With respect to telegraphy, he was not concerned with the simple relationship between a sender and receiver facing each other like A and B but with multiple ramifications that started from one point: a kind of spider's web. In the lecture, we hear about the relationship between the "governing centre" and the "extremities" and at the same time about the formation of a "will" in the center, which is "executed" on the periphery.[22] This is not just an allusion to the geopolitical order of the Prussian state and a concrete allusion to the optical

FIGURE 28. Telégrafo/Telegraph, drawing by Francisco de Goya (ca. 1824–28). © Madrid, Museo Nacional del Prado.

telegraph line from Berlin to Coblenz but also an allusion to the structure of the human organism, especially the arrangement of brain and nerves. As precedent for this approach, Helmholtz cites the example of the Roman consul Menenius Agrippa, who was said to have managed to calm down a group of insurgent plebeians "by wisely likening" the body to the state.[23]

At the same time, he brings the effects of this structuring, including its calming effect, to bear in the way he structures his talk. Immediately following the introduction of the analogy of state telegraphy and human body, he names the "principal question" his research sought to answer:

> In the transmission of such messages [*Nachrichten*], is a measurable time necessary for the ends of the nerves to communicate to the brain the impression made upon them; and on the other hand, is time required for the conveyance of the commands of the will from the brain to a distant muscle?[24]

This description literally turns on its head what Helmholtz's procedure had actually consisted in: the multilayered drifting of research machines ever since the days of the Potsdam frog drawings. What, in December 1850, had been placed at the beginning of the report as a capital problem has become manifest on the level of laboratory practice as an unpredictable offshoot of a process with unknown outcome.

Thanks to this process, in the course of a whole year, a highly complex assemblage of organic and mechanic, electric and magnetic, optical and handwritten components was tried out and explored: first with regard to muscles, then to nerves, and finally to the brain as well. In later years, Helmholtz was to admit: "In my memoirs I have . . . not given the reader an account of my wanderings, but I have described the beaten path on which he can now reach the summit"—the goal of the experimental investigation—"without trouble."[25] His time experiments no doubt mark one of the passages in which this shift in emphasis, from "wandering" to "the beaten path," becomes palpable.

Yet discourses have their own consistency, and the procedure for answering the alleged principal question accordingly appears as planned and calculated:

> A slight electric shock is given to a man at a certain portion of the skin, and he is directed the moment he feels the stroke to make a certain motion as quickly as he possibly can, with the hands or with the teeth, by which the time-measuring current is interrupted.[26]

What is measured this way is the "interim between the stimulation and the muscle's action" or the "sum" of the individual processes taking place "from the stimulation of the sensory nerves till the moving of the muscle."[27] In concrete numbers, this interval amounts to a total of 0.125 to 0.200 seconds.

What exactly happens in this time span remains unclear. Helmholtz merely emphasizes that the numerical values of the measurements within a given series of observations coincide "sufficiently well." "It is found, for example, that a *message from the great toe* arrives about 1/30th of a second later than from the ear or the face."[28] Here too, the *how* of measuring determines the *what* of what is measured.

In an additional step, Helmholtz then presents the relatively low divergence of individual results as a promising precondition for deriving the duration of partial aspects of the reaction from the "sums" obtained. Here, Helmholtz on the one hand presumes "that the duration of the processes of perceiving and willing in the brain does not depend upon the place on the skin at which the impression is made."[29] On the other, he is aware that the measured sums of time vary and depend "chiefly upon the degree of attention; it varies also at different times in the case of the same person." Only when the test subjects' attention was "fixed" did the sums become "very regular."[30] If this condition was met, it was possible to assume that in the stimulation of different points on the skin across the human body "only the first member" of the reaction process changed: the propagation of the stimulation in the sensory nerves.

These reflections are still being developed when the analogy with telegraphy comes in: it makes it easier to dissect the time sums measured into physiological elements. Thus, Helmholtz explained, a part of the time that passes between stimulation and reaction is used up by "transmitting the message [*Botschaft*]"—that is, by the propagation of the stimulus in the sensory nerves. Another part of the time, which he presumed to be equally great, is said to be needed for the message to "travel" from the brain through the motor nerves to the muscles.[31]

From his frog experiments, he then supplied the insight that one-hundredth of a second passes before the muscle, having "received" the message, even enters into action. This was the interim that was to become, in the "Second Note," the *temps perdu*. Yet that is where he left it in the Königsberg lecture. Accordingly, the remainder had to be that time "which passes while the brain is transferring the message [*Depesche*] received through the sensory nerves to the motor ones."[32] This was the cerebral tenth of a second, which he presented in his lecture as a result in addition to the propagation speed of the nerve stimulation: a remainder left over after the subtraction of all other "intervals" presumably traversed.

The whale

In the conclusion of his Königsberg lecture, Helmholtz sets out to determine the place and status his "human time measurements" occupy in human experience. In this remarkable passage, the physiological fact of cerebral and nervous interims acquires an almost poetic character. As little doubt as there could be for Helmholtz that the propagation of stimuli in nerves and brain took place at comparatively low speeds, he was aware of the obvious inaccessibility of this fact both to common sense and to attentive self-observation. Just after introducing his "principal question," Helmholtz had remarked about the time taken up by sensations coming in from the body's periphery and by the will exercised in the center that "in our own case we have never perhaps experienced anything similar."[33] He now returned to this point.

To be sure, "daily experience" showed "that, owing to the time required for the transmission of sound, we hear after we see"[34]—think of a thunderstorm or a cannon being fired. It was also known, Helmholtz explained, that light and sound phenomena that follow on one another in quick succession have a tendency to melt "into one." A "glowing coal" quickly moved in a circle appears as a "continuous circle of fire."[35]

These were indications of the peculiar imprecision of the human perception of time. Yet this hardly made the fact that both the propagation of nerve stimulations and the transmission of messages of the will took up time *within* the human body any more palpable, intuitive, or concrete. The "sum" of time situated between sensing a stimulation and executing a movement might be experienceable, for example as a reaction that is more or less rapid, more or less successful. But the brain and nerve times Helmholtz had measured as partial sections of the reaction process withdrew from such experience; they could by definition not be experienced because they continually passed outside of human experiencing, inexorably and unnoticed. In Helmholtz's terms: The interims settled in a region in which "the expression of our own experience can give us no information"[36]—a kind of physiological a priori accessible only via precision time measurements and, in this respect, closely tied up with another scene, another *Schauplatz*: the laboratory.

Given this difficulty, it hardly comes as a surprise that, at the end of the lecture, Helmholtz once more draws on an experiment for illustration. This time, however, he does so not with respect to quantitative results but in regard to a qualitative experience from the day-to-day work in the laboratory, a report on how obstacles in conducting an experiment are encountered. Only now does Helmholtz return to his experiments on human motor nerves, those experiments that had been the concrete starting point of his human time measurements. As he now admits, these experiments had led to "no exact results," but they suggested "other interesting relations connected with the subject,"[37] in particular the relation of the involuntary to the voluntary, of the physical-unconscious to the psychic-conscious.

The twitching of the forearm muscles that had been prompted by the electric stimulation of the meridian nerve in the first series of human time experiments and the hand movements that resulted from it were in no way to be understood as reflex actions. They resembled actions that were withdrawn from the control of the will *through time*. It turned out, in fact, "that these motions are totally independent of the influence of the will, because the will, informed of the shocks by the sensible nerves, cannot exert itself sufficiently soon upon the muscles."

This had become especially evident in a series of experiments in which the test subjects' task had been to maintain their hand in the very position it had twitched into as a result of an electric stimulation of the upper arm. Maintaining this position had turned out to be impossible because the hand "fell back very speedily"[38] into its original position: too speedily for the will to contravene this movement. The paradoxical result was a maintaining that could not be maintained. The declared intention to no longer move the hand, to fixate it, went nowhere. Instead, one observed a rapid repetition of the first hand movement: not a controlled fixation but a short, unavoidable dithering, as if the body had for a moment escaped the will and immediately been recaptured.

The artificiality of the example undermines its plausibility. Who among Helmholtz's listeners had had similar lab experiences? That is why, in the last paragraph, Helmholtz returns once more to the structure of the human body, not, this time, to illustrate a physiological fact but, on the contrary, to illustrate its lack of illustrative capacity—and perhaps also to give an explanation for the communication difficulties that had come up since the "Preliminary Report."

Helmholtz explained that the temporal relations of brain and nerve functions he explored were not accessible to everyday consciousness because they were anchored deep in human anatomy. This seemed reassuring: "Happily the distances are short which have to be traversed by our sensuous perceptions before they reach the brain, otherwise our self-consciousness would lag far behind the present."[39] By way of conclusion, a look at comparative anatomy was to make clear why exactly we were lucky:

> With an ordinary whale the case is perhaps more dubious; for in all probability the animal does not feel a wound near its tail until a second after it has been inflicted, and requires another second to send the command to the tail to defend itself.[40]

Thanks simply to its anatomy, the human being could defend itself more quickly.

But for the physiologist, this lucky circumstance changed nothing about the fact that human self-consciousness de facto still lagged after the present—not at too great a distance but still in a way that could be measured: in the execution of arbitrary actions, for example, by a cerebral tenth of a second. Nor did the anatomical circumstance change anything about the fact that, in Helmholtz's view, the involuntary was not merely reflexive but was also motoric, activated by stimulations, which raced ahead of what was sensed and what was willed. For everyday people, he admitted, knowing about such "interesting situations" made no difference. They could rest assured that we do not notice self-consciousness lagging behind the present and that we "are therefore unprejudiced in our practical interest."[41]

Alexander von Humboldt confirmed this point in his own way. After reading the printed text of Helmholtz's lecture, he expressed his amusement and related an anecdote. He wrote to du Bois-Reymond:

> The whale . . . reminds me of the joke about Chateaubriand's extremely tall former secretary, Monsieur de Valéry, when he became Court Librarian. "He never needs a ladder to take down a book, and when Monsieur de Valéry is wading through the mud, it takes two weeks for the snuffles to rise to this head."[42]

The interim Helmholtz had measured with such technical expenditure is thus associated with and subordinated to the seemingly familiar phenomenon of incubation. Rising from the cold wet feet to the nose, it in a way nonetheless remains a message from the big toe.

In such associations, it becomes clear that Helmholtz's psychophysiological time measurements led to a remarkably abstract kind of knowledge. Although it concerned the basics of human experience, this knowledge remained irrevocably linked to the scene of the lab. In the end, the time measurements provided no further insight into the phenomenon they at the same time established in such precise fashion: the propagation of stimulations through the nerves and the brain. They shed no further light on the functioning or "mechanics" of nerve action, of the nervous agent. They only made it more plausible that these were materially based processes, not the effect of an imponderable *fluidum* like electricity, for example. Nor could the interims Helmholtz had tracked down be experienced, even though they were literally passing at every moment. The temporal gap Helmholtz had detected between nerve stimulus and muscle action opened a window through which the brain and the nervous system radiating out from it became visible, yet they remained as dark and inaccessible as a black sun. And in this regard, the lecture given before the Königsberg Physical-Economic Society on December 13, 1850, elicited more "wonder" than "comprehension"—an effect he had hoped for, Helmholtz claimed in a letter to du Bois-Reymond, who, on this point, however, was of a decidedly different opinion.[43]

By this point, Helmholtz's human time measurements have revealed themselves to be a scientific *event* in the emphatic sense, as an invention that manifests itself in laboratory things and laboratory signs, in experimental actions and reactions, and, at the same time, produces a new virtual world, a future that comprehends an incompletable succession of variations of the same event.

Following up on Deleuze's reflections, Joseph Vogl has claimed that the event exists "only in a specific time-in-between, a specific space-in-between," in the empty times of hesitation and waiting and the equally empty spaces that consist of disconnected places and positions not occupied.[44] With regard to the material and semiotic practice of Helmholtz's human time measurements, one could add that it is the interims, the times-in-between that turn into the event, that become an emphatic indication of a precarious and productive research machine made up of heterogeneous components, whose effects cannot be simply conceptualized and which therefore also incite further experiments, further measurements.

Such was not Helmholtz's declared intention. Two days after the lecture, on December 15, 1850, he wrote a "Message for the Berlin Physical Society Concerning Experiments on the Propagation Speed of Stimuli in the Sensory Nerves of the Human Being," as if he wanted to close the parenthesis that he had opened in January of that same year with the "Preliminary Report." This "Message" presented the scientific content of the human time measurements in short form—without any analogy, any explanatory example whatsoever. Despite all problems with the reception of the "Preliminary Report," he picked up on this report in the new "Message" by giving a short and, once again, not illustrative description.

On four manuscript pages, Helmholtz first sketched how he had applied "Pouillet's method" to measure, in human subjects, the "interval between the stimulation and the movement setting in."[45] Concretely, the task was to measure the sensation of an electric shock on a delimited point on the skin (neck, face, finger, calf, etc.) by means of a "movement of the hand or the teeth" that was to follow "as quickly as possible" upon the perception of this sensation. Again, he professed his surprise at how little the values he found diverged:

> Given the fixed attention of the individual subject to the experiment, the measurement results obtained in a series of observations lasting from 1 to 1½

hours coincided with one another so well one might not expect it. While the measured values as a whole range from 0.12 to 0.20 seconds, the probable error in each individual observation turns out to be, in successful series, 0.003 sec., [the error] of their median [*Mittelzahl*] 0.001 sec. or even less.[46]

Again, Helmholtz pointed out the decisive role attention plays. He notes that "slightest sickness" and "tiredness," above all, lead to "very irregular numbers." And then the results obtained were concretized in precise figures: "From the difference between large toe and sacrum resulted a speed of 62.1 ± 6.7 M[eters per second], between finger and neck a speed of 61.0 ± 5.1 M[eters per second]."[47]

In the conclusion, Helmholtz used these numbers, as he had done in the popular lecture he had given two days earlier, as the basis for deriving the time needed "for the processes of perception and volition in the brain": 0.10 seconds. With regard to the particular demands placed on the test subjects in these time measurements, he then added:

> Of course this only refers to the cases presented here, where the decision as to what is to happen has been taken and one only waits for the signal. If at the time of perceiving the signal, one's thoughts are occupied with something else and the mind only now recalls what movement is to be executed, it takes much longer.[48]

This highlighted the decisive role of focused attention in precise human time measurements—and it sounded a leitmotif of later studies of this kind, namely, the test persons' concentration in the lab.

In contrast to the "Preliminary Report," however, Helmholtz's message about the propagation speed in human sensory nerves remained unpublished. All that was published from it in his lifetime was a short notice in the *Advancements in Physics*, the journal of the Physical Society. In a list of "original research" presented by members of the society in the sessions conducted in 1850, there is the entry "20 Dec. *Helmholtz*. On the propagation speed of nerve stimulation in the sensory nerves."[49] That it was the sensory nerves *of the human being* that were concerned was not explicitly mentioned.

SEVEN

The Return of the Line

> Like me, the line is searching without knowing what it's searching for.
>
> —HENRI MICHAUX[1]

What followed for Helmholtz was a further drifting of the experimental process and a return to the method of the curves. The drift this time led from psychophysiology to physics. In the human time measurements, it had been necessary to work with "strong apparatuses," that is, with powerful resistors that kept the electric shocks so weak that they were "just barely noticeable."[2] This drew attention to a problem that had not until then been taken into account: When exactly does an induction shock take effect physiologically? Does it act as if immediately—at the same high speed as electricity—or is there a delay, due to the coils or some other factor? Helmholtz put it as follows: In his human time measurements there was the danger that "time differences" be ascribed to "nervous actions" when in fact they belonged to the electrical current.[3]

This problem led him to undertake a separate study that occupied him from the winter of 1850–51 onward. At the beginning of April 1851, he sent the communication that resulted from this work, "On the Course and Duration of Electrical Currents Induced by Fluctuations of Currents," to du Bois-Reymond to have it—according to the familiar pattern— presented before the Berlin Academy and the Physical Society and to then have it published. Everything that followed also adhered to the familiar scheme.

Shortly afterward, on May 8, 1851, a short excerpt from the work was read in the Berlin Academy, not by Johannes Müller this time but by Johann Christian Poggendorff. The detailed account was then published in a scientific journal: not, however, in the *Archiv für Anatomie, Physiologie und wissenschaftliche Medicin*, edited by Müller, but in the *Annalen der Physik und Chemie*, edited by Poggendorff.[4] Du Bois-Reymond, too, reacted the familiar way, with criticism similar to that of the "Preliminary Report." Given the once again largely hermetic nature of Helmholtz's text, he had confronted once more the necessity of "reinventing":

> I also have to confess to you that I am not at all happy with your presentation.
> I have read through your treatise and the abstract several times without
> understanding what you had actually done and how you had done it. Finally,
> I invented the method myself and only then did I gradually understand your
> presentation.[5]

Then he adds, almost angrily:

> You must—don't take this the wrong way—be more diligent in abstracting
> from your own standpoint of knowledge and put yourself where those stand
> who do not yet know what it is about and what you want to explain to them.[6]

Du Bois-Reymond thus once more demanded an "inductive" kind of presentation, one that led with logical and historical consistency from not-knowing to knowing.

The pattern of Helmholtz's answer seems familiar, too. Slightly disappointed but ultimately unimpressed, he writes back to Berlin:

> As for the presentation in the article, it has been a particular effort for me
> this time and I finally thought I could be satisfied with it. But you're right,
> the more one corrects, the more difficult it often becomes to understand.
> The topic, incidentally, was a bad one to treat of.[7]

In other words, if scientific things are as complicated as they are, that is not the experimenter's fault.

Helmholtz's return to the curve method took place in parallel with this drifting. This return can be understood as picking up once again on the Potsdam frog drawing machine and the "simple apparatus" for physiological recordings he had still used even in his early Königsberg days. But it also recalls Matteucci's deployment of the curve method for measuring the velocity of muscle contractions.

The reasons for this return explicitly named by Helmholtz are, of course, of another kind: epistemic reasons, but technical ones as well. And temporal considerations play a role, too: for example, the time needed for conducting individual series of experiments. As Helmholtz had explained in his December 1850 lecture before the Königsberg Physical-Economic Society, his aim after the preliminary conclusion of his time measurement experiments on frogs' motor nerves had been to conduct equivalent experiments on warm-blooded organisms as well, in order to leave the "simplest cases" of animal physiology behind him.[8]

The human time measurements begun in spring 1850 had been a first foray in this direction. As we have seen, they were based on including the entire human body in the experimental setup. They had the real advantage of working with longer nerve distances and the possible disadvantage of now having to work with conscious reactions, no longer just with involuntary contractions. It would have been another foray in the same direction to have conducted experiments with samples of isolated nerves and muscles taken from warm-blooded animals—rabbits, for example. Compared to cold-blooded frog samples, however, such samples did not keep nearly as well: extended "series of measurements which occupy from two to three hours,"[9] such as were required by the Pouillet method, could therefore not be performed. The method of curves seemed to offer a remedy here because it led to results more quickly.

However, one recording of the muscle contraction was not enough to measure the propagation speed of the nerve stimulation. Instead, two distinct contractions triggered from different points on the nerve had to be graphically recorded in such a way that it was possible to deduce this speed from comparing the contractions and from comparing the correlating time sums. Helmholtz had already described the basic principle of this deployment of curve drawing in his Königsberg lecture:

> If two different points of a nerve be stimulated successively, and if the moments of stimulation coincide exactly with one and the same position of the point upon the rotating cylinder, two congruent curves are produced, which, however, appear moved towards each other in a horizontal direction. The magnitude of the displacement corresponds to the time of propagation in the length of nerve between the two points of stimulation.[10]

Taken together, this staggered writing of curves above or into each other saved a considerable amount of time in experimental labor. And this economy "may perhaps be made use of" for working with samples from warm-blooded animals, as Helmholtz added, rather circumspectly, at the end of the lecture.[11] The epistemic motivation of moving on from frog physiology to the more complex facts of the physiology of warm-blooded organisms had, at this point, clearly receded in favor of technical and temporal aspects. The curve method was faster, that was the main thing.

There were other, implicit reasons for returning to the curve method. We may again call them aesthetic reasons. In contrast to the complex and abstract character of the electromagnetic time measurements according to Pouillet's procedure, the muscle-movement's self-recording on a rotating cylinder promised to yield results that were much more intuitive than numbers and tables. After the partly reticent, partly dismissive reactions his number-centered communications had elicited in Berlin as in Paris, Helmholtz seemed to be willing to give credence to this promise—despite the fundamental criticism he had articulated concerning the friction potential of the curve method in the "Measurements" just a few weeks earlier. In the meantime, however, du Bois-Reymond had reported in detail on his stay in Paris and sketched the problems he had had in communicating Helmholtz's results. Once more, it is temporal factors that are relevant—in this case the time needed to show a physiological fact.

This was the main reason Helmholtz wanted to use a drawing machine "to place before everyone's eyes, in 5 minutes, the fact of the duration of propagation in nerves in an experiment."[12] One single gesture, intelligible to anyone, was to suffice to overcome the reservations of the academic public: placing staggered curves *sous les yeux* as a time-saver argument. The technical preconditions had been met since September 1850, when Helmholtz had asked Egbert Rekoss, the Königsberg instrument maker with whom he also collaborated in building the ophthalmoscope (the other main achievement of Helmholtz's work in Königsberg),[13] to manufacture the parts he needed for retooling his research machine: a glass cylinder, a precision-mounted lever, a steel pin, and so on.

In December of that same year, he was able to run the "first explorative experiments on frogs"[14] with his newly equipped machine. The results thus obtained, however, did "not yet" possess "the same degree of precision and consistence" as those he had conducted according to Pouillet's method.[15] The curve method had a long way to go before it could provide reliable results by itself, either automatically or mechanically. Yet apparently, Helmholtz mastered these difficulties in the following weeks. In spring 1851, he wrote to du Bois-Reymond that he was "adjusting [the] rotating cylinder for human beings" to be able to conduct "time-measuring physiological experiments."[16] Like Pouillet's method, it seems, the curve method had turned out to be so promising that it was to be extended from frogs to other model organisms.

Yet it was several more months before Helmholtz began to publicly report on his use of the curve method in detail—first in September 1851 in the short "Second Note on the Propagation Speed of the Nervous Agent" that du Bois-Reymond translated (see chapters 1 and 2), then, about a year later, in a detailed and illustrated article in the *Archiv für Anatomie, Physiologie und wissenschaftliche Medicin*, entitled "Measurements Concerning the Propagation Speed of Irritations in the Nerves (Second Series)." The experiments brought up in both these publications, the "Second Note" and the "Measurements (Second Series)" were part of a strategy that remained a rather defensive one. They did not aim at studying warm-blooded organisms (neither rabbits nor human beings) but continued to work with nerve-muscle samples taken from frogs. In addition, the new series of experiments aimed less at measuring time than at clarifying what had been properly measured in the earlier experiments. Helmholtz repeated earlier experiments, albeit with small differences. He conducted his chronoscopic experiment on frogs' motor nerves once more on the register of chronography yet subordinated this "once more" to the purposes of clarification. In this regard, his deployment of the curve method was motivated aesthetically, not primarily epistemically. And that is also why the detailed article did not live up to the promises of its title. As in the "Second Note," Helmholtz did not, properly speaking, report on "measurements on the propagation speed of stimulations in the nerve"—but on demonstrations in physiology.[17]

The machine behind the curves

If we follow the detailed description given in the "Measurements (Second Series)," we can once again describe the changes in Helmholtz's research machine in terms of an exit and an entry: the galvanometer was discarded along with the reading telescope and replaced with a "drawing cylinder" that was attached to the side of the frog frame. The frame with the suspended nerve-muscle sample was retained, as was the first electrical circuit with which the sample was stimulated. Also discarded were the steel frame with the contact plates and the mercury interface, as well as the shallow dish for the weights at the lower end of the steel frame. Now instead, attached to the lower end of the frog sample, was a ladder-shaped lever that inclined horizontally toward the glass cylinder. Affixed to the side of this lever was a thin steel pin that served as stylus or "drawing point" (*zeichnende Spitze*)[18] for scratching the curves into the layer of soot applied to the cylinder. As for the cylinder, it was, as already mentioned, "a piece from a thick, near-cylindrical champagne glass cut-off to fit" and "very precisely" polished by Rekoss.[19]

The glass cylinder was driven by a clockwork with a cone pendulum. Clockworks with "springing gear" were obviously not suitable for the purposes of curve recording. What was needed was a clockwork that turned in an "uninterruptedly uniform" manner. As Helmholtz critically remarks, however, "practical mechanics" had not yet managed to "strictly solve" this problem. The way out of the impasse consisted in at least regulating the clockwork in a certain way. Such regulation was to ensure that possible fluctuations in the motion of the clockwork could take place only slowly. To this end, "a heavy flywheel lined with lead" was affixed to the axle supporting the drawing cylinder: "Given the great inertia of this disk, the velocity of its rotation changes only very slowly when the clockwork's driving forces become somewhat greater or smaller."[20] It was not, therefore, the cone pendulum attached at the bottom of the clockwork that was used for regulation. This pendulum, Helmholtz explained, only served as a means for determining the number of the cylinder's rotations, usually six rotations per second. That was the drawing machine's operating speed, and he could read it off the distance between the rotating balls from one another (Figure 29).

The concrete fabrication of the curves took place as follows. A "perpendicular lever arm" and a "thumb" that could close a contact and effect a stimulation of the muscle were attached to the flywheel below the cylinder. This switch-like device had been described, albeit rather tersely, in the "Second Note," as a "protruding cog."[21] It now served, first, to mark the "point of the cylinder" that corresponded to "the moment of the stimulation" at a certain point on the nerve. The cylinder thus initially stood still. Then it was slowly turned by hand until the so-called "thumb" on the flywheel touched the contact lever and the muscle contracted. The result was "a simple vertical line" that marked nothing but the moment of the stimulation.[22]

Then, the stylus was moved away from the cylinder, which also moved the "thumb" away from the lever, and the cylinder was set in motion until it had reached the desired rotation speed: "As soon as one notices the balls beginning to separate, the drawing can be executed."[23] Again the stylus was placed against the cylinder, again the "thumb" tapped against the lever, and again a contraction was triggered. This time, the course of the contraction was recorded on the cylinder.

Then Helmholtz removed the stylus from the cylinder again, stopped the cylinder, and marked the curve that had been drawn with a check mark, "in order to be able to distinguish it later with certainty from the second curve yet to be executed."[24] For applying these marks, incidentally, Helmholtz used a couching needle, an instrument he was familiar with from physiological optics or from ophthalmology.

After marking the curve, another point of the nerve was switched into the circuit and the entire procedure was repeated—under time pressure.

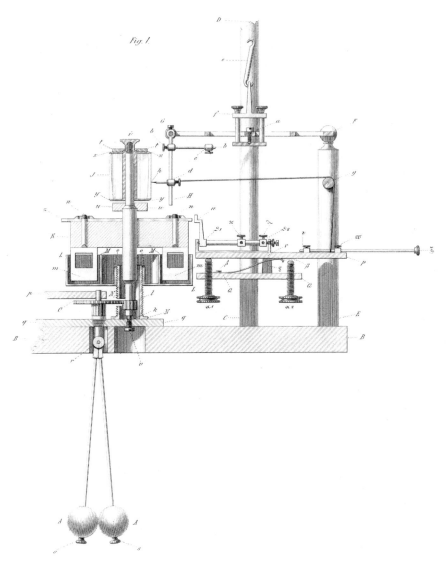

FIGURE 29. Illustration of central components of Helmholtz's curve drawing machine (1852). The focus is on the glass cylinder, the steel tip, and the metal frame—a detailed view of the myograph later depicted by Marey as well (see above, fig. 7). This drawing, unlike Marey's, includes the flying wheel (hatched) and the cone pendulum below the recording drum. The nerve-muscle sample and the suspension frame are not depicted. Reproduced from Hermann Helmholtz, "Messungen über Fortpflanzungsgeschwindigkeit der Reizung in den Nerven (Zweite Reihe)," *Archiv für Anatomie, Physiologie und wissenschaftliche Medicin* (1852): 199–216, table.

For the more quickly the second recording took place, "the more certainly will one find the muscle's stimulatability to not have changed noticeably the second time around, which is an essential condition for the success of the experiment."[25] Only if it could be guaranteed that the only variable changed was the stimulation point on the nerve could one be certain that one had in fact performed a measurement of time.

After the conclusion of the second curve recording, the curves obtained needed to be detached from the glass cylinder and fixed. As in the unpublished supplementary paragraph of the "Second Note," Helmholtz describes the way he proceeded here in detail in the "Measurements (Second Series)" as well. To fix the curves, he rolled the cylinder across "a plate of fish glue that had been breathed on."

> Breathing onto the glue plate makes it a little sticky and latch on to the cylinder's soot such that the drawing is taken off the cylindrical and onto the plane surface. The sheet of glue can be placed, soot side down, on a wet sheet of white paper to which it sticks. The curves then appear white on a black ground and are very clearly visible.[26]

Based on the curve drawings thus obtained, Helmholtz then produced by hand the master drawings for the reproductions on the plate that accompanied the "Measurements (Second Series)."

We may plausibly suppose that the curve drawings he used there were similar to those preserved at the Académie des sciences. If so, then the changes the curves underwent on their way into print become quite clear. They intervene, so to speak, where the work of translation and implementation of the unpublished "Explication des épreuves" had left off: at a "double curve," drawn in isolation and ideal form, marked with Greek letters and Arabic numerals and inserted into a kind of small coordinate system (see chapter 2).

Like the hand drawings included with the manuscript of the "Second Note," the curves that were finally printed were enlarged as well, even though their width on the printed plate is surprisingly close to the width of the preserved transparencies. Helmholtz explains that during copying, the curves' "vertical heights" had been "doubled"—"for reasons of perspicuity."[27] An even more important change consisted in reversing the direction of the curves. The direction in which the recorded contraction process now unfolded corresponded to the usual reading direction from left to right, which is also why the arrows, still needed on the boards of the Académie curves, were absent. Not until this point, then, did "the conventions of the Latin script" also determine the physiological curves' kind of graphical representation: "two-dimensional application onto paper, running direction from left to right."[28]

Something else stands out in the curve figures in the "Measurements (Second Series)": there is no indication of time on the abscissa. Despite its title, the publication is only secondarily concerned with *measuring* time. As Helmholtz writes explicitly at the beginning of his discussion of the curves, he did not use the curves for measuring but "for *representing* the propagation speed in the nerve."[29] The decisive function of this "representation" consisted in guaranteeing that the only difference between successive experiments was that the point at which the nerve had been stimulated had changed. To clarify this point, the "Second Note" emphasized first that the recordings of successive curves were perfectly congruent.[30] The "Measurements (Second Series)" take the exact inverse approach. Here, Helmholtz first shows that the curve recording does *not* take place with perfect congruence, especially if one disregards another temporal factor, the muscle's "diminished stimulatability."[31]

This was another aspect of the new method's temporality. First, it was clearly faster, on the whole, than measuring time with Pouillet's procedure; second, it was able to demonstrate the physiological fact in question more quickly; yet, third, it also had to be performed as rapidly as possible to guarantee the uniformity of the curves. If one waited too long between two experiments or conducted too many experiments between the first and the second recorded experiment, one did not obtain congruent curves. What one then obtained, what then became visible, as the first double curve Helmholtz comments on demonstrates, were two staggered curves with similar bents but of clearly different heights and widths (Figure 30), a difference "due to the gradual diminishing of the sample's stimulatability."[32] At this point, therefore, not only does time become palpable, a further drifting announces itself, a drifting, however, that would not lead to bigger projects for several decades: toward the experimental physiology of *fatigue*.

After this presentation of (curve) difference followed the representation of identity, albeit a shifted identity. The next figure on the plate of the "Measurements (Second Series)" shows, next to the first, a second curve that, "in respect to strength, duration, and progression of the individual stages of the contraction [was] very much the same" (Figure 31).[33] What these curves running in parallel make clear is that one of the contractions had indeed occurred later than the other, that the shifted drawing could therefore only be attributed to the change in where the nerve had been stimulated: "Since in both cases the adjustments of the apparatus and the mechanic forces of the muscle were the very same, the delay of the effect in the one case can only have been caused by the longer propagation in the nerve."[34] "Longer propagation" is here to be understood in both the

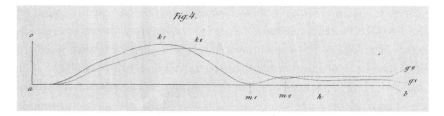

FIGURE 30. Noncongruent double curves of a worn-out nerve-muscle sample (1852). Reproduced from Hermann Helmholtz, "Messungen über Fortpflanzungsgeschwindigkeit der Reizung in den Nerven (Zweite Reihe)," *Archiv für Anatomie, Physiologie und wissenschaftliche Medicin* (1852): 199–216, table.

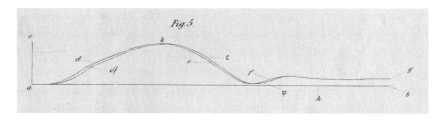

FIGURE 31. Congruent double curves, shifted because of the stimulation being applied in different places (1852). Reproduced from Hermann Helmholtz, "Messungen über Fortpflanzungsgeschwindigkeit der Reizung in den Nerven (Zweite Reihe)," *Archiv für Anatomie, Physiologie und wissenschaftliche Medicin* (1852): 199–216, table.

temporal and the spatial sense, as a longer distance to run through in the nerve and a larger interval this takes up.

This makes clear what was at stake in demonstrating the fact that the repetition differed in only one precisely circumscribed point: the fundamental structure of Helmholtz's time-measuring experiments and the preconditions for applying his variation and subtraction method. In Helmholtz's view, the decisive contribution of his method was "that in every single drawing of two curves that belong together, one can immediately tell from their shape whether the muscle has worked uniformly in both cases."

Two more illustrations were supplied to support this argument (Figure 32). They reproduce the check marks applied to the glass cylinder by hand with a couching needle. Here, as in the unpublished "Explications des épreuves," they were meant to signal that, independent of whether the first point on the nerve to be stimulated was located far away from the muscle or, inversely, the first point to be stimulated was located close to the muscle, in both cases the uniform curves shifted in relation to one another. The shift thus showed nothing less than the contraction's delay when the stimulation took place at the more distant point. Once again, the laboriously produced equivalence of experimental conditions turned out to be the empty center around which the recording practice of the "new method" organized itself.

At the same time, the problem of measurement recedes into the background. Helmholtz remarks almost casually: "As far as the absolute value of the propagation speed is concerned, the horizontal distances of the two curves cannot be measured with very great precision."[35] Unlike the first, explorative curve published in 1850 in the "Measurements," the abscissae in the double curves printed two years later are not marked with any units of time. This, however, does not keep Helmholtz from stating that the values of the velocity measured are "about as great" as the ones obtained earlier by means of Pouillet's procedure.

This statement is based on a survey of the curves that also took the cylinder's circumference (85.7 millimeters, about 3.4 inches) and the number of its rotations into account. A distance of one millimeter in the curve drawing thus corresponded to 1/514 of a second. Taking into consider-

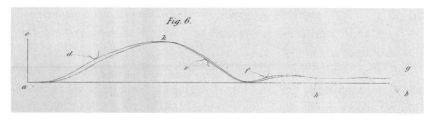

FIGURE 32. Congruent double curves, in which a check mark marks whether the muscle was first stimulated at a point closer or more distant on the nerve (1852). Reproduced from Hermann Helmholtz, "Messungen über Fortpflanzungsgeschwindigkeit der Reizung in den Nerven (Zweite Reihe)," *Archiv für Anatomie, Physiologie und wissenschaftliche Medicin* (1852): 199–216, table.

ation as well the distance between the two points at which the nerve had been stimulated in each of the recordings (namely, 53 millimeters, about 2 inches), the propagation speed in the nerve could be deduced: 27.25 meters per second (about 89.5 feet per second).[36]

Helmholtz does not comment on the reliability of this way of measuring time. As had already been the case in the "Second Note," the number stands, without further comment, beside the results of the earlier measurements according to Pouillet's procedure. What follows is a reminder, nothing more: "The most probable value in the earlier experiments was 26.4 meters."[37]

The depiction of the four double curves on the plate that comes with the "Measurements (Second Series)" was to remain Helmholtz's only publication of the curves. In the spring of 1852, he once again wrote to du Bois-Reymond that he had "arranged the new apparatus for the human time measurements." He evidently still planned at this point to deploy the curve method for his studies on and with human research objects. But such experiments were not conducted in the following years and such curves were never published.

In 1854, there is a last message from Königsberg based on the curve method. But this message, too—titled "On the Velocity of Some Processes in Muscles and Nerves," and read, once again through du Bois-Reymond's intercession, in the Berlin Academy of Sciences and published in its *Bericht über die zur Bekanntmachung geeigneten Verhandlungen*—refers to experiments on frogs and contains neither results of precise time measurements nor illustrations of any kind.[38]

The four printed pages instead offer descriptions of different phases in the contraction of the muscle under changing circumstances: the intervention of secondary twitching, the stimulation of electrical shocks administered in quick succession, and the triggering of reflex movements. On this last point, for example, we read that "the reflected twitching sets in only after relatively long intervals"[39]—a formulation that is characteristic of the qualitative-descriptive tone of the message as a whole.

What is remarkable is something else. In a manner similar to the "Second Note," the first section of this message summarizes the results obtained thus far on the temporal course of muscle activity in a terminology that had not been used before. Helmholtz, to be sure, does not speak of "lost time" when it comes to the delayed contraction. The two texts are not *that* close. But we find evocative terms such as "interval of the latent stimulation" or "period of the latent stimulation"—phrases that pick up on an expression current in the 1830s and 1840s in the works of Johannes Müller, for example, or Hermann Lotze. There, the term "latent" served, among other purposes, to describe the field of consciousness, for example in the case of several simultaneously present but unfocused representations or in discussions of the relationship of matter on the one hand and the principle of life or the soul on the other.[40] In the following years and decades, the phenomenon of latency in nervous action, also referred to as the "refractory period," became a major topic of neurophysiological research.

The curves' afterlife

As for Helmholtz, he turned to other topics. With the new ophthalmoscope in hand, he mainly focused in the 1850s on his interest in physiological optics (in 1856, the first volume of his *Handbook* on the topic appeared), then on physiological acoustics. Not until the late 1860s, after his move from Bonn to Heidelberg, was he to return to his human time measurements—without, however, publishing any curves on that occasion either. It was thus left to others to show the first curves that refer to the time relations anchored in the human brain and nervous system.

After a remarkable pause, an "interim" of almost fifteen years, a student of Helmholtz's, Rudolf Schelske, was the first to publish such example curves in his 1864 "New Measurements of the Propagation Speed of Stimuli in Human Nerves."[41] At that time, Schelske was not working in Heidelberg, with Helmholtz, but in Utrecht, in ophthalmology. As he himself explains, his major goal in executing these new human time measurements was to close a technical gap. Until then, time measurements on humans had only been undertaken according to Pouillet's method. According to Schelske, it was "desirable" to take such measurements also using "the second, graphical, method"—on the one hand because it was impossible to reproduce "in its former shape" and "as a whole" the experimental setup according to Pouillet that Helmholtz had initially used, "since parts of it were left behind in Königsberg," and on the other hand because techniques of curve drawing had continued to develop since the early 1850s.

Supported by new registration apparatuses, especially from astronomy, Schelske was no longer concerned with representing curves but with curve measurements of the propagation speed of nerve stimulations. The technical step he took to achieve this purpose is as simple as it is convincing. Besides the recording of stimuli and reactions he also used electromagnetism to draw the second-by-second signals of a metronome on a rotating drum. With a "small microscope," these pinnacle-shaped "double curves," consisting each of a stimulus-reaction curve and a time curve, could be compared with precision, and the time spans in question could be measured to within one-thousandth of a second (Figure 33).[42]

Other physiologists at the time—Donders, for example, and Marey after him—made use of tuning forks to scratch a precisely known number of vibrations per seconds onto a rotating recording surface next to traces of stimuli and reactions. This procedure, too, seems to have been developed in Paris in the 1840s, by the Austrian-born physicist Guillaume or Wilhelm Wertheim.[43] While drawing next to each other stimuli and reactions on the one hand and time signals on the other was the simplest way of turning the curves into measurements of time, this very kind of parallel action also

FIGURE 33. Curve recording for the purpose of time measurement (1864). The upper curve is the recording of a signal of seconds given by a metronome. A microscope served to divide this signal into smaller portions of time. The lower curve shows the duration between the stimulus (c) and the reaction registered as well as, subsequently, the preparation of the next experiment, which begins at the next stimulus. Reproduced from Rudolf Schelske, "Neue Messungen der Fortpflanzungsgeschwindigkeit des Reizes in den menschlichen Nerven," *Archiv für Anatomie, Physiologie und wissenschaftliche Medicin* (1864): 151–73, here 161, fig. 5.

rendered the curves superfluous, in a way. The curves had served their purpose once they had been counted or surveyed and a time value been noted down. Numbers sufficed—the curves as curves were no longer needed.

Donders's way of proceeding makes this clear in exemplary fashion. In his studies on the physiological time of psychological processes, he preserved numerous data sheets on which he had noted down the values of the tuning fork vibrations he had counted in each case, but he did not preserve a single curve drawing. Tabulating the time signals had rendered these recordings obsolete.[44]

The Helmholtz curves had an afterlife nonetheless. Marey, especially, showed them in the publications in which he also elaborated on the concept of the "temps perdu," in *Du mouvement dans les fonctions de la vie* as well as in *La machine animale* and *La méthode graphique*. He did not, to be sure, use in these works Helmholtz's curves in their first (printed) form, by reprinting the plate published in the "Measurements (Second Series)" of 1852. They were published in a slightly modified form that Marey at one point took from a popular lecture on the time experiments conducted by Helmholtz, Schelske, and Donders. It was du Bois-Reymond who had given this lecture and shown the corresponding curves in the Royal Institution in London in 1866.[45] These curves, which were shifted sideways but otherwise congruent and without time-measuring accompanying curves, reappear two years later in Marey's book on movement in the life functions. Not unlike Helmholtz, Marey describes them as a "representation," inspired by Helmholtz's publications, of the delay of successive muscle twitches that come about depending on the distance between the nerve points stimulated (Figure 34).[46]

He nonetheless did not hesitate to supplement—and thereby also to replace—this *répresentation* with measurements. Not unlike Donders, Marey made these measurements by way of parallel recordings of muscle contractions and tuning fork vibrations, albeit not on a rotating cylinder but on a rotating disk covered with a layer of soot.[47] The curves thereby appear bent, curved in turn—comparable to what must have been the case in Matteucci's early time-measurement experiments. Marey does not deploy rotating cylinders for "measuring the velocity of the nervous agent" and for imprinting straightened-out drawings until the 1870s (Figure 35). That, however, remains an isolated case. In general, Marey was more concerned with the "striking form" (*forme saisissante*) of his curves than with precision measurements.[48]

The reprises and transformations of the Helmholtz curves in Schelske, Donders, Marey, and others, however, form only one part of the iconography

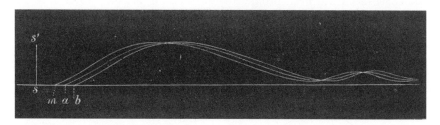

FIGURE 34. Illustration of a variant of the Helmholtz curves in Marey (1868). "Representing, following Helmholtz, the successive delay of muscle jolts according to the point at which the stimulus is applied: – s, electrical stimulus. – m, jolt of the muscle stimulated directly. – a, jolt produced by a stimulation of the nerve close to the muscle. – b, jolt produced by stimulating the nerve far from the muscle." Reproduced from Etienne-Jules Marey, *Du mouvement dans les fonctions de la vie* (Paris: Baillière, 1868), 418.

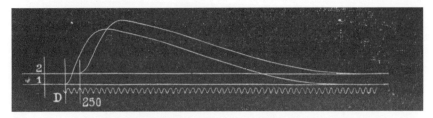

FIGURE 35. "Measurement of the speed of the nervous agent" according to
Marey (1878). Two curves of frog muscle contractions provoked by stimulations
at different points on the nerve are accompanied by the curve of a tuning fork
vibrating 250 times a second. The *temps perdu* is marked with particular clarity
here because the time measurements set in only after it has passed, at point D.
Reproduced from Etienne-Jules Marey, *La méthode graphique dans les sciences
expérimentales et particulièrement en physiologie et en médecine* (Paris: Masson,
1878), 147.

the time experiments conducted by the Königsberg physiologist entailed.
The other part leads us back to the research machine behind the "Preliminary Report" and the "Measurements," back to the comparatively abstract
and complicated applications of Pouillet's method with which Helmholtz
had undertaken precise time measurements on prepared frog samples and
on whole human beings from October 1849 onward. We thus move from
the curves that had been produced since the fall of 1850 with a drawing
machine back to the lines that trace a past seeing machine from the spring
of that year and outline new, future research machines. We return from
representations to the diagram.

In the text of the "Preliminary Report," Helmholtz had represented the
experimental setup of frog frame, galvano-chronometer, and reading-
telescope in an "excessively obscure" way, as du Bois-Reymond had remarked.
Even in the plate accompanying the ninety pages of the "Measurements," he
had only shown partial aspects of the setup. These aspects were detailed
differently: the frame and bell jar were pictured in extreme detail, the
galvano-chronometer only rudimentarily, the reading-telescope not at all.
None of his publications present the experimental setup in a complete,
clearly structured way. In the aftermath, both the specific way of focusing
of Helmholtz's published drawings and the effective complicatedness of the

experimental setup he used led to images being produced and distributed that showed his heterogeneous research machine in its entirety—with more or less intrusive changes.

The explicit purpose of those images was to visualize a machine whose achievements and functions were deemed exemplary of the culture of precision in the emerging field of experimental physiology. Without saying so, however, these visualizations also reflected the difficulties that struck Schelske, Donders, and other physiologists as they set out to reproduce Helmholtz's machine measurement results under changed technical conditions. Yet in precisely these uncertainties of reconstruction and construction, the images in question convey the latitude that was to contribute to the continuation and further evolution of Helmholtz's time experiments in other places and at later points in time. In this respect, they can be seen as approximating the diagram defined at first only by experimental practices in Königsberg. They show a machine made up of organic and nonorganic, physiological and physical components, a machine that seems to showcase both the precariousness of its functioning and its astonishing productivity.

The images, too, are remarkably heterogeneous. They oscillate between stylized representations of an electrical circuit and naturalized depictions of instruments and test samples. They merge schema and drawing, juxtapose overview and aspect. The first image is included in Carl Ludwig's 1852 *Textbook on Human Physiology* (Figure 36). Ludwig shows a rectangular arrangement in which only the nerve-muscle sample and the galvanometer needle retain something representational, whereas all other components of the experimental setup are turned into lines and arrows, letters and numbers.

Somewhat less schematic is the figure in Carl Kuhn's *Manual of Applied Electricity Theory* (Figure 37), which, not unlike Schelske's essay, is published at a remarkable fifteen years' remove from Helmholtz's publications. The materiality of the prepared frog sample and the galvanometer comes through, also that of the induction coil, the rocker switch, and the suspended steel frame. As already in Ludwig, however, Kuhn's image does not show the telescope setup necessary for precision in reading off the measured values— presumably because one did not want to draw the human experimenter into a representation that was complex enough as it was.

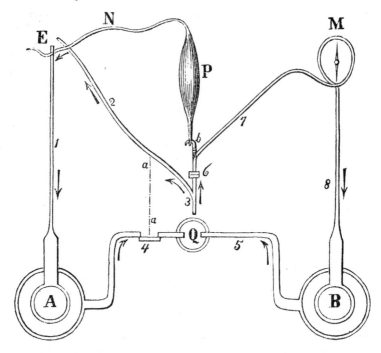

FIGURE 36. Ludwig's illustration of the setup used by Helmholtz for precision measurements of the propagation of nerve stimulations (1852). (A) and (B) are galvanic elements, (M) the needle of a galvanometer, (P) a frog muscle, (E N) the corresponding nerve. The numbers 1 through 8 denote conductive wires, (a) and (b) isolating rods, (Q) is the little cup of mercury that allows for the permanent closure of the contact. The time measurements are taken by varying the position of wire 2 on the nerve (E N): from the current point to one close to the muscle. Reproduced from Carl Ludwig, *Lehrbuch der Physiologie des Menschen* (Heidelberg: Winter, 1852–56), vol. 1, 115.

The same is true for the image that can be found in the already mentioned lecture du Bois-Reymond gave in 1866 before the Royal Institution and that Marey reproduced as well. This lecture, "On the Time Required for the Transmission of Volition and Sensation through the Nerves," reads like an elaborate reprise of du Bois-Reymond's attempt, first made in Paris in 1850, at conveying the significance of Helmholtz's time experiments— and thus the significance of the Berlin school of organic physics—to an

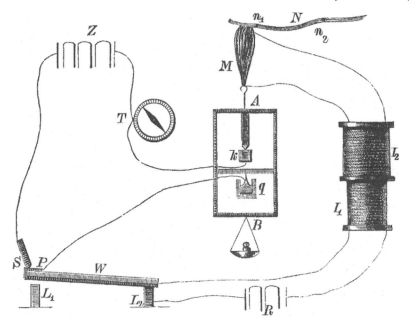

FIGURE 37. Kuhn's illustration of the setup used by Helmholtz for precision measurements of the propagation of nerve stimulations (1866). This scheme clearly distinguishes two circuits. In the first circuit, the battery (R), connected to the inductor coils (I$_1$) and (I$_2$), served to stimulate the prepared nerve-muscle sample (N/M). The second circuit connected the galvanometer (T) with the battery (Z). The switch (S/P) made it possible to close both circuits at the same time. The setup in the frame (A) made sure that the time-measuring current was permanently interrupted after the contraction of the frog muscle. Affixing an electrode at different points on the nerve (n1, n2) allowed Helmholtz to obtain the time the nerve needed to conduct the stimulation. Reproduced from Carl Kuhn, *Handbuch der angewandten Elektricitätslehre, mit besonderer Berücksichtigung der theoretischen Grundlagen* (Leipzig: Voss, 1866), 1193.

audience of physiologists as well as nonphysiologists. Du Bois-Reymond, too, shows Helmholtz's experimental setup as a combination of heterogeneous parts—steel frame and muscle, wire and nerve—but adds focalizations on the points of contact between electricity and physiology, which are described in detail in his lecture as well. In the years that follow, comparable images that bring out the assemblage character, the almost cyborg-like quality of Helmholtz's time experiments, find their way into the textbooks and manuals of physiology.[49]

Yet du Bois-Reymond did not want to leave it at spoken words and printed paper. In London he also presented Helmholtz's time-measuring experiments *in actu*. This was the big difference from his first public reference to these experiments in Paris in 1850. The image in the published text of the London lecture is thus also to be understood as representing a demonstration experiment carried out before the members of the Royal Institution. This adds to its interest because it does not simply show an experimental setup but also signals that the experiment has itself become a means of showing. In fact, Helmholtz's time-measuring experiments were to remain a standard in du Bois-Reymond's pedagogical repertoire— side by side with the light beams projected via mirror galvanometers and the presentation of "twitching telegraphs," "frog pistols," and other demonstration machines.[50]

In the lecture hall of the institute du Bois-Reymond opened on Dorotheenstraße in Berlin in 1877, the time-measuring experiments thus entered into the context of object lessons, in which "seeing for oneself" and "ascertaining for oneself" had become the most important task of prospective physiologists.[51] Whereas Marey, in *La méthode graphique*, showed the curves, which he had had drawn to measure the propagation speed of the nervous agent, and in this context criticized Pouillet's method, which Helmholtz had initially used, as imprecise,[52] du Bois-Reymond in his teaching insisted on the significance of the heterogeneous research machine that had been the first to allow for precise measurements of this velocity. Countless students drew sketches of this machine into their notebooks—ironically by taking recourse to a graphical method: freehand drawing (Figure 38).

In his teaching as in his lecture at the Royal Institution, du Bois-Reymond emphasized the unsettling and interesting moment in Helmholtz's time experiments. He stressed their "beauty and the high scientific value" and took pains—as Helmholtz did himself—to exemplify their status and significance in human experience.[53] To do so, he picked up on the example of the harpooned whale who cannot begin to defend himself until "one whole second" after he has been struck. Yet the Berlin physiologist did

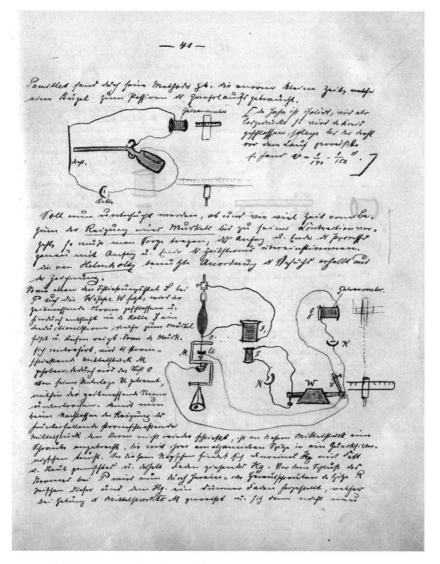

not want to have to rely completely on the illustrative power of comparative anatomy.

With another example, he went into the thick of the increasingly technicalized and scientificized life-world of the nineteenth century. To give his audience a succinct impression of the particularities of psychophysiological time relations, he turned to a discussion of velocities that had been established in the 1860s for other "bodies" or "agents": the velocity of electrical currents that, according to Wheatstone's experiments, amounted to 464,000,000 meters per second, the speed of light (around 300,000,000 meters per second, according to Fizeau), then the speed of sound in iron, in water, and in the air, that of a cannon ball, of wind, and the speed of a railroad engine (27 meters per second), as well as that of a rock being thrown up 24.5 meters by a human hand (21.9 meters per second).[54]

As if to conjure these more or less rapid velocities, du Bois-Reymond lists them in the spatial form of a table. This tableau, which could serve as the cornerstone for a comparative theory of speed in the nineteenth century, appears as a unique anticipation of the overviews in manuals of practical photography that much later—around 1890—listed the basic conditions of *genuine* "focused" or "clear-cut snapshots." Beside distance and exposure time, the velocity of the process photographed was one of these conditions. Was the picture to be taken of a mountaineer or a racehorse, the growth of a fingernail or a comet tail?[55]

Although du Bois-Reymond does not, at this point, explicitly refer to instantaneous photography, his representation and demonstration of Helmholtz's time experiments is concerned with a similar fixing technique, a procedure that leads to a "freeze-frame."[56] To explain this procedure in more detail, the physiologist jumps from biology to technology, from the whale to the railroad engine, and from there to another cyborg, the railroad engineer.

The perspective he sketches in the process is unusual, to say the least. He assumes an observer posted along the railroad tracks who would be able to see, at the same time, a moving train and the propagation of a stimulation in a nerve.

> Suppose the engine-driver on the locomotive of an express-train running a mile a minute holds his arm extended towards the tender, and moves his fingers, then the nervous agent in the motor fibres of his arm will rest in space or nearly so, because its motion is destroyed by that of the train.[57]

The comparative table of velocities seems to lend credibility to this claim: the nerve stimulus moves at 26 to 30 meters per second from the engine in the direction of the tender. The train, in turn, moves at 27 meters a second (thus about 60 miles an hour) in the opposite direction. An external observer who could see both movements at the same time would thus witness the movement of stimulation in the nerve being arrested.

Does this seem improbable or difficult to accept? Du Bois-Reymond explains it again, slightly differently: "And the same thing will happen with the nervous agent in the sensory fibres of the fireman's arm, if, standing on the tender, he should burn his hand at the locomotive."[58] In this case, too, the propagation of the stimulation, which takes place in the direction opposite to that of the train, would be as if canceled out by the movement of the train.

The hypothetical observer at this point thus sees two locomotion workers in whom nerve stimulations are halted at full speed, once on the way from the center of the body to its periphery, once in the opposite direction. This redoubles, A.D. 1866, a polar inertia, as if to emphasize the very trick exemplified by the Helmholtz curves: producing a deceleration with the means of acceleration.[59] Surrounded by generalized progress, these curves do in fact mark, in exemplary fashion, a circumscribed standstill, a pause, an interim.

Conclusion

> Some used to say that art in a period of speed and haste would be brief, like the people before the war who predicted that it would be over quickly. The railway was thus supposed to have killed contemplative thought, and it was vain to long for the days of the stage-coach, but now the automobile fulfils their function and once again sets the tourists down in front of abandoned churches.
>
> —MARCEL PROUST[1]

Proust and Helmholtz never met. When the "Reich Chancellor of Sciences" died in Berlin in 1894, the budding author was all of twenty-three years old. Proust never went to Berlin, and Helmholtz didn't want to go to Paris. He preferred traveling to England or the United States. Proust probably never read a single line of Helmholtz. It seems that the discourse of "temps perdu" was delivered to the future author of the *Recherche* via Marey.

Where Helmholtz and Proust meet nevertheless is in a creative use of dynamization technologies for distancing themselves from social and cultural modernity within its very center. The physiologist and the author make use of telegraphy and of newspapers to brake and to delay. The one thereby defines an experimental framework within which visualizations and measurements of psychophysiological time are elaborated to this day. The other created a novel that was meant to be like a gothic cathedral before which one can stand forever.

Different as their ways of proceeding were, Helmholtz and Proust agree in their conclusion that human behavior and experience is at its core a discontinuous process, and both men experience, even if each in a different sense, that the semiotic world of modernity is an arbitrary world.

That is why their search for lost time remains also a search for lost signs. Deleuze considered this to constitute the unity of the *Recherche*. In this unique work, he writes, signs are not the objects of abstract knowledge but of temporal apprenticeship, an enduring *apprentissage*. What the apprentice learns, however, during this time of being taught, is what it means to learn in the first place: "To learn is first of all to consider a substance, an object, a being as if it emitted signs to be deciphered, interpreted."[2]

A similar kind of unity marks the texts and images with which Helmholtz renders account of his time experiments. Here, too, the physiologist appears as apprentice, as an interpreter of the muscles and nerves, the instruments and wires he manipulates at his laboratory bench.

At the same time, that is the difference. For Helmholtz does not only observe, he also intervenes, he seizes, he assembles, he produces. Before the experimenter sets out to decipher, he enciphers. In the laboratory, deciphering is always preceded by enciphering. Prior to interpretation, there is always another interpretation. Put differently, even before he sets out to calculate and interpret his results, Helmholtz entices the materials, objects, and entities in his laboratory to emit additional and, one hopes, unexpected signs.

"There is no apprentice who is not 'the Egyptologist' of something," Deleuze writes about the *Recherche*.[3] This is as true of Marcel the narrator as for Proust the author. Yet it is also true of the lab and life scientist Helmholtz—except that he is not just an Egyptologist, he is also *the Egyptian of the experiment*.

Commentary on the life sciences in the nineteenth century has insisted time and again that Darwin had made it impossible to understand time merely as a uniform chronology of dates independent of each other. On this reading, the *Origin of Species* irrevocably turned time into

> the movement . . . by which the universe has become what it is, a process of development, a change from the most elementary to the most complex; in short an 'evolution' born of the internal sequence of transformations.[4]

Yet it was not just the differentiation of organisms into different species that in the nineteenth century transformed the temporal from an external parameter into an internal operator. Thanks to Helmholtz's time experiments, time became a critical, an incisive and decisive, factor in the living existence of organisms as well.

Around 1850, the organic physicist in Königsberg assembled an astonishing research machine that consisted of frog muscles and galvanometers, batteries and human test subjects, telescopes, wooden frames, and bell jars. This machine allowed him to show that in animals as in human beings, the coordination of muscles, nerves, and the brain follows a temporal pattern that is characterized, not by simultaneity or continuity, but by gaps, interims, *temps perdus.*

Despite the difficulties that opposed communicating and spreading these results, there can no longer be any doubt that individual organisms' respective relationships with their environment expose them to an ongoing withdrawal of time. From the point of view of the experimental life sciences, individuals have to come too late to be able to have a consciousness of punctuality at all. In order to advance, they are forced to lag behind themselves.

The Helmholtz curves stand for the insight into this physiological and psychological fact. But they do not illustrate it. They do not make this fact visible. What then do they show? To return to the beginning of this book, what phenomenon or what process do they render intuitive?

The answer I have given is that the Helmholtz curves show a fundamental precondition of psychophysiological time measurement experiments. *Two successive executions of the same experiment in which the only parameter changed was the stimulated nerve point while all others remained the same or were kept the same as much as possible* led to a lateral shift in the recorded contraction curves. In other words, the Helmholtz curves do not serve to "persuasively represent" the "results" Helmholtz had obtained in his research with the Pouillet method. The curves were no doubt easier to take in than comparatively sober rows of numbers, and to that extent, they were also more convincing, at least to those for whom facility was a decisive criterion.

With his curves, however, Helmholtz did not represent results but preconditions. The curves do not bring results into a clear arrangement.

Ultimately, they do not even stand for a specific physiological or psychological process. The Helmholtz curves show a specific kind of experiment, a procedure, a method. The image they offer is an image of experimental dependence.

Taking up a distinction that Helmholtz makes on the basis of his time experiments, we can articulate this point in another way. The Helmholtz curves, we may say, function more as *symbols* or signs than they do as *images*. For a sign—Helmholtz didn't need Saussure to see this—"need not have any kind of similarity at all with what it is the sign of."[5] Indeed, there is no correspondence in form, perspective, or color between these curves and the propagation speed of stimulations in nerves. They are certainly indexical signs, imprints, or "inscriptions," to use the terms of both Marey and Latour. But an intuitive similarity between these curves and the movement they record does not exist.

A muscle contraction—not to speak of the nerve stimulation and its velocity—does not look like the wave- and dune-shaped lines on the reverse side of a film of glue that has preserved its remarkable transparency during all its years in the archive. Precisely because they are indices, *empreintes*, this kind of similarity does not exist. What inferences can we make about what an animal looks like from the traces it leaves in the sand? What does a fingerprint tell us about the appearance of a human being?

Such relationships of reference may be one of the reasons why Helmholtz, in discussing his curves, never speaks of immediate perception or intuitiveness (*Anschaulichkeit*). The lines and arcs he drew with the help of his machine (and the machine with his) were meant to provide a faster and better "overview" (*Überblick*) of the phenomenon in question. There were not, to draw on Marey one last time, merely meant to be *inscriptions*, they also functioned as *representations*.[6]

This does not mean, however, that these curves do not have an aesthetic of their own. Quite the contrary. This aesthetic just turns out to be much more modernist than has been assumed. It has very little to do with Romanticism's facial lines, with anatomical body shapes or sublime landscapes. Instead, it instantiates an arbitrariness and openness that is deeply inscribed in the modern mode of existence.

As representations of a laboratory procedure, these curves do indeed lead us back to the machines that were used in physiological and psychological laboratories to produce, receive, and spread the most varied kinds of signs. From the drawing of curves we are thus led to the heterogeneous assemblages with which time could be precisely measured, using electricity and light, nervous actions and muscle twitching, to swiftly and easily handled telegraph keys as well as to inert, slowly deflecting galvanometer needles.

I would like to call the spaces in which the work on these time-sign machines took place (and still takes place) *laboratory fractals*. They are spaces that repeat, within a given laboratory, the architectural and technological structures by means of which scientific enterprises separate and insulate themselves from their everyday surroundings in order to reconnect with them through other channels: via letters, journals, and newspapers, but also via wires, tubes, and similar infrastructures.

For this reason, too, does every sign lose time at every point, when it is produced as much as when it is received, when it is sent off as much as when it is delivered. This accounts, at least in part, for the precipitation that comes with the successful production of a decipherable sign—and for the historical forgetfulness that accompanies communications pertaining to this sign.

As a result, not just space, but time too appears peculiarly fractalized. The temporal relations that dominate on the exterior correspond to those that manifest in the interior. A synchronization of body parts continues and develops the synchronization of clocks. On the edge of an extended network of simultaneities, time researchers lie in wait for the shortest of messages, signals, and information that can be communicated across the living body. What in the internal world appears as interim and *temps perdu* becomes manifest in the external world as congestion, as hesitation and perplexity—and vice versa.

Within such constellations, even the difference between scientific and artistic experiment is determined in terms of time. Proust, at least, thought so: "An impression is for the writer what an experiment is for the scientist, except that for the scientist the work of intelligence precedes it, and for the writer it comes afterwards."[7]

Helmholtz confirms and generalizes this perspective by connecting experimentation and life. For him, experiments are not only the central epistemic tool of all sciences of life; they are deeply inscribed in the body, i.e. the unconscious of the modern subject:

> The same great importance which experiment has for the certainty of our scientific convictions, it has also for the unconscious inductions of the perceptions of our senses. It is only by voluntarily bringing our organs of sense in various relations to the objects that we learn to be sure as to our judgments of the causes of our sensations. This kind of experimentation begins in earliest youth and continues all through life without interruption.[8]

In the life-laboratory of the nineteenth century, seeing time was more precise than writing time. Yet chronoscopy was much more complicated and more abstract than chronography. For Helmholtz, this was not equivalent to saying that it also was more artificial. Precisely because of its artificiality did chronoscopy approximate the living process of human perception. In the end, the Helmholtz curves refer us back to that experimental process as well.

1844

December 23: Pouillet tells the members of the Académie des sciences in Paris how "extremely short intervals of time" can be measured with the help of a galvanometer.

1845

January 14: The Berlin Physical Society is founded. Its founding members include Wilhelm Beetz, Emil du Bois-Reymond, Ernst Brücke, Wilhelm Heintz, Gustav Karsten, and Hermann Knoblauch.

January 20: Reacting to Pouillet's presentation, Breguet introduces to the Académie the "apparatus for measuring the velocity of a projectile at different points of its trajectory" that he has developed with Konstantinov.

March 7: At a meeting of the Berlin Physical Society, du Bois-Reymond suggests applying Pouillet's method to research muscle and nerve activity.

April 14: In the Académie, a letter from Matteucci to Alexander von Humboldt about "new studies on frog electricity" is read.

May 20: Du Bois-Reymond drafts a letter to von Humboldt, in which he claims priority over Matteucci. He does not send the letter.

May 26: Before the Paris Academy, Charles Wheatstone reports on electromagnetic chronoscopes. He accuses Breguet (and Konstantinov) of plagiarism.

October 3: Werner Siemens joins the Pouillet-Breguet-Wheatstone debate. Before the Berlin Physical Society, he gives a lecture, "On the Application of Electrical Sparks in the Measurement of Velocities," in which he claims priority in matters of electromagnetic measurements of short intervals of time for the Prussian Artillery Inspection Commission.

1846

December 12: Carl Ludwig, in Marburg, uses a kymograph to record circulation and breathing movements of horses and dogs in the form of curves ("first stammering of heart and breast").

1847

June 10: Before the Royal Society in London, Michael Faraday reads a message from Matteucci (signed in Pisa in February), in which Matteucci reports on experiments on measuring the duration of muscle contractions with the help of electromagnetism and the recording of curves.

July: The first volume of the Berlin Physical Society's yearbook, *Advancements in Physics in the Year 1845*, is published.

July 21: Helmholtz to du Bois-Reymond: "I would like it if you could get the essay by Matteucci for me."

July 23: Helmholtz gives his lecture "On the Conservation of Force" before the Berlin Physical Society.

1848

July 18: In a letter to his fiancée, Olga von Velten, Helmholtz mentions a "frog drawing machine" that makes it possible to draw "very fine and regular" curves of muscle movements.

1849

May 17: Alexander von Humboldt obtains publication in the Paris Academy's *Comptes rendus* of a description of du Bois-Reymond's electrophysiological experiment, in which a test subject deflects a galvanometer's needle by flexing his arm muscles.

End of August: Shortly after their wedding, Hermann and Olga Helmholtz settle in Königsberg.

October: Helmholtz reports to du Bois-Reymond that he has taken up once more his studies on muscle activity, according to a "whole new method." This is the electromagnetically based procedure of measuring time according to Pouillet.

1850

January 15: Helmholtz signs the "Preliminary Report" and sends it to Berlin, to du Bois-Reymond, von Humboldt, and Johannes Müller.

January 21: Müller reads Helmholtz's "Preliminary Report" before the Prussian Academy of Sciences.

February 1: Du Bois-Reymond reads the "Preliminary Report" before the Physical Society. The report is subsequently published in three journals: first the proceedings of the Prussian Academy (*Bericht über die zur Bekanntmachung geeigneten Verhandlungen der Königlich Preussischen Akademie der Wissenschaften zu Berlin*), then in the *Archiv für Anatomie, Physiologie und wissenschaftliche Medicin*, and finally in the *Annalen der Physik und Chemie*.

February 25: Via von Humboldt, the content of the "Preliminary Report," in a translation by du Bois-Reymond, is communicated to the Académie des sciences in Paris.

February 27: *Le National* runs a short piece on the report by "M. Helmottz."

March 15: Du Bois-Reymond goes to Paris to present to the Paris Academy and to publicly confront Matteucci.

March 25: Du Bois-Reymond presents his studies to the Paris Academy, but he also reports on Helmholtz's time experiments.

March 30: Du Bois-Reymond presents his studies to the Société philomatique and, once again, reports on Helmholtz's time experiments.

End of March/beginning of April: Helmholtz begins his "human time measurements," the extension of the Pouillet method from frog samples to human bodies.

March 29 to June 9: Hermann and Olga Helmholtz's laboratory logbook.

May 21: Du Bois-Reymond returns to Berlin.

June 28: Helmholtz finishes the manuscript of the "Measurements," which describes the application of the Pouillet method in detail.

Middle of September: Helmholtz orders an apparatus for the auto-registration of muscle movements from Egbert Rekoss in Königsberg.

December 13: Before the Königsberg Physical-Economic Society, Helmholtz reports on the new methods of precision time measurement, his application of these methods in experiments on prepared frog samples and human beings, and on first test runs with the new instrument for the recording of curves.

December 15: Helmholtz writes a note to the Berlin Physical Society, in which he reports on his experiments on the propagation speed of stimulations in the sensory nerves of human beings.

December 20: Helmholtz's note on the "human time measurements" is read before the Physical Society in Berlin.

1851

April 11: Helmholtz sends his treatise on the course and duration of induction currents to du Bois-Reymond.

May 8: Poggendorff reads an excerpt from the "induction essay" in the Prussian Academy of Sciences.

August 6 to September 28: Helmholtz visits physiological research institutions and German-speaking universities and presents his frog curves "everywhere."

September 1: Helmholtz's report on the use of the curve method is presented to the Paris Académie des sciences. He includes two curve drawings with the written report (see figures 1 and 2) as well as an explanatory text, the "Explication des épreuves [Explanation of the Proofs]" (figures 9 and 10).

September 8: The Paris Academy sets up a commission to study Helmholtz's two reports. Beside Pouillet, its members are the physiologists Pierre Flourens and François Magendie.

1852

June 28: Helmholtz gives his habilitation lecture in Königsberg on the nature of human sense perception.

1854

June 12: Du Bois-Reymond presents the Prussian Academy with Helmholtz's report on the speed of some processes in muscles and nerves.

1855

Excerpts from the "Measurements," edited by the physicist Marcel Verdet, are published under the title "Mémoire sur la contraction des muscles de la vie animale et sur la vitesse de propagation de l'action nerveuse."

1856

July: Together with his family, Helmholtz leaves Königsberg. He takes up a position at the University of Bonn.

INTRODUCTION

1. Gaston Bachelard, *The Dialectic of Duration*, trans. Mary McAllester Jones (Manchester: Clinamen, 2000), 92 [modified].

2. Marta Braun, *Picturing Time: The Work of Etienne-Jules Marey (1830–1904)* (Chicago: University of Chicago Press, 1992).

3. Daniel Kehlmann, *Measuring the World*, trans. Carol Browne Janeway (London: Quercus, 2007), 233. On this highly suggestive account, see Ottmar Ette, *Alexander von Humboldt und die Globalisierung: Das Mobile des Wissens* (Frankfurt and Leipzig: Insel, 2009), 302–18.

4. See Paul F. Cranefield's classic paper, "The Organic Physics of 1847 and the Biophysics of Today," *Journal of the History of Medicine* 12 (1957): 407–23.

5. Bachelard, *Dialectic of Duration*, 77: "The history of the laboratory phenomenon is very precisely that of its measurement." From the perspective of the history of science, see Thomas Kuhn, "The Function of Measurement in Modern Physical Science," *The Essential Tension: Selected Studies in Scientific Tradition and Change* (Chicago: University of Chicago Press, 1977), 178–224; and the partly iconoclastic chapter on "Measurement" in Ian Hacking, *Representing and Intervening: Introductory Topics in the Philosophy of Natural Science* (Cambridge: Cambridge University Press, 1983), 233–45.

6. Hebbel E. Hoff and Leslie A. Geddes, "Graphic Registration before Ludwig: The Antecedents of the Kymograph," *Isis* 50, no. 159 (1959): 5–21.

7. Etienne-Jules Marey, *La méthode graphique dans les sciences expérimentales et particulièrement en physiologie et en médecine* (Paris: Masson, 1878), III.

8. Ibid.

9. Carl Ludwig, *Lehrbuch der Physiologie des Menschen*, vol. 1 (Heidelberg: Winter, 1852), 114.

10. Charles Marx, "Le neurone," *Physiologie*, ed. Charles Kayser (Paris: Flammarion, 1969), vol. 2, 7–285, here 16.

11. Kathryn M. Olesko and Frederic L. Holmes, "The Images of Precision: Helmholtz and the Graphical Method in Physiology," *The Values of Precision*, ed. Norton Wise (Princeton, NJ: Princeton University Press, 1994), 198–221, here 198; and "Experiment, Quantification, and Discovery: Helmholtz's Early Physiological Researches, 1843–1850," *Hermann von Helmholtz and the Foundations of Nineteenth-Century Science*, ed. David Cahan (Berkeley: University of California Press, 1993), 50–108.

12. On this point, see my extensive study *Hirn und Zeit: Die Geschichte eines Experiments, 1800–1950* (Berlin: Matthes & Seitz, in press).

13. Gilles Deleuze, *Cinema 1: The Movement-Image*, trans. Hugh Tomlinson and Barbara Habberjam (Minneapolis: University of Minnesota Press, 1986), 56.

14. Ibid.

15. Gilles Deleuze, *Cinema 2: The Time-Image*, trans. Hugh Tomlinson and Robert Galeta (Minneapolis: University of Minnesota Press, 1989), 211. See also Henri Bergson, *Matter and Memory*, trans. Nancy M. Paul and W. Scott Palmer (New York: Macmillan, 1929), 30. Concerning the interstice, see also Joseph Vogl, *On Tarrying*, trans. Helmut Müller-Sievers (London: Seagull, 2011), as well as Bernhard J. Dotzler and Henning Schmidgen, eds., *Parasiten und Sirenen: Zwischenräume als Orte der materiellen Wissensproduktion* (Bielefeld: Transcript, 2008).

16. Justus von Liebig, *Familiar Letters on Chemistry*, 3rd rev. ed. (London: Taylor, Walton & Maberly, 1851), 272. On analytic experimentation, see Frederic L. Holmes, "The Intake-Output Method of Quantification in Physiology," *Historical Studies in the Physical and Biological Sciences* 17 (1987): 235–70; and John V. Pickstone, *Ways of Knowing: A New History of Science, Technology and Medicine* (Chicago: University of Chicago Press, 2001), 83–134.

17. Bergson, *Matter and Memory*, 30.

18. Hermann Helmholtz to Olga von Velten, July 18, 1847, *Letters of Hermann von Helmholtz to his Wife*, ed. Richard L. Kremer (Stuttgart: Steiner, 1990), 43.

19. Sensory nerves [*sensible Nerven*], in the wide sense, as distinct from the narrow sense [*sensorische Nerven*], denoting nerves that participate in sense perception.

20. Peter Galison, *Image and Logic: The Material Culture of Microphysics* (Chicago: University of Chicago Press, 1997).

21. Wolfgang Schivelbusch, *The Railway Journey: The Industrialization and Perception of Time and Space in the Nineteenth Century*, trans. Anselm Hollo

(Berkeley: University of California Press, 1986); Stephen Kern, *The Culture of Time and Space, 1880–1918* (Cambridge: Harvard University Press, 2003); Paul Virilio, *Negative Horizon: An Essay in Dromoscopy*, trans. Michael Degener (London: Continuum, 2005); Peter Weibel, *Die Beschleunigung der Bilder in der Chronokratie* (Bern: Benteli, 1987); and Edna Duffy, *The Speed Handbook: Velocity, Pleasure, Modernism* (Durham, NC: Duke University Press, 2009).

22. Robert M. Brain and M. Norton Wise, "Muscles and Engines: Indicator Diagrams and Helmholtz's Graphical Methods," *Universalgenie Helmholtz: Rückblick nach 100 Jahren*, ed. Lorenz Krüger (Berlin: Akademie, 1994), 124–45. More generally, see Horst Bredekamp, *The Lure of Antiquity and the Cult of the Machine: The Kunstkammer and the Evolution of Nature, Art and Technology*, trans. Allison Brown (Princeton, NJ: Wiener, 1995).

23. Timothy Lenoir, "Helmholtz and the Materiality of Communication," *Osiris* 9 (1994): 185–207.

24. The crucial text in this respect is Claude Pouillet, "Note sur un moyen de mesurer des intervalles de temps extrémement courts, comme la durée du choc des corps élastiques, celle du débandement des ressorts, de l'inflammation de la poudre, etc.; et sur un moyen nouveau de comparer les intensités des courants électriques, soit permanents, soit instantanés," *Comptes rendus hebdomadaires des séances de l'Académie des sciences* 19 (1844): 1384–89.

25. See Christian Bonah, *Les sciences physiologiques en Europe: Analyses comparées du XIXe siècle* (Paris: Vrin, 1995). On the role of journals in the life sciences of the nineteenth century, see James A. Secord, *Victorian Sensation: The Extraordinary Publication, Reception, and Secret Authorship of "Vestiges of the Natural History of Creation"* (Chicago: University of Chicago Press, 2001).

26. The classical sociological perspective on this issue is developed by Robert K. Merton, "The Reward System of Science (1957)," *On Social Structure and Science*, ed. Piotr Sztompka (Chicago: University of Chicago Press, 1996), 286–304. More recent approaches figure in Mario Biagioli and Peter Galison, eds., *Scientific Authorship: Credit and Intellectual Property in Science* (New York: Routledge, 2003).

27. See the remarkable charts and maps in Johann Christian Poggendorff, *Lebenslinien zur Geschichte der exacten Wissenschaften seit Wiederherstellung derselben* (Berlin: Duncker, 1853).

28. Walter Benjamin, "The Paris of the Second Empire in Baudelaire," trans. Harry Zohn, *Selected Writings*, Volume 4: *1938–1940*, ed. Howard Eiland and Michael W. Jennings (Cambridge: Harvard University Press, 2003), 3–92, here 31.

29. Jonathan Crary, *Suspensions of Perception: Attention, Spectacle and Modern Culture* (Cambridge: MIT Press, 1999). On the instantaneous in

photography, see Raymond Bellour, ed. *Le temps d'un mouvement: Aventures et mésaventures de l'instant photographique* (Paris: Centre national de la photographie 1986); François Albera, Marta Braun, and André Gaudreault, eds., *Arrêt sur image: Fragmentation du temps/Stop Motion: Fragmentation of Time* (Lausanne: Payot, 2002); and Phillip Prodger, ed., *Time Stands Still: Muybridge and the Instantaneous Photography Movement* (New York: Cantor Center for Visual Arts at Stanford University/Oxford University Press, 2003).

30. On cutting and cropping in photography see Rosalind Krauss, "Stieglitz/'Equivalents,'" *October* 11 (1979): 129–40. On the frame in painting see Georg Simmel, "The Picture Frame: An Aesthetic Study," trans. Mark Ritter, *Theory, Culture & Society* 11, no. 1 (February 1994), 11–17.

31. Helga Nowotny, *Time: The Modern and Postmodern Experience*, trans. Neville Plaice (Cambridge: Polity, 1996), 80. Remarkable ideas concerning a time and rhythm analysis of scientific practice can be found in Bachelard, *Dialectic of Duration*, 67–80.

32. Concerning the experimental process as an interplay between difference and repetition, see Hans-Jörg Rheinberger, *Toward a History of Epistemic Things: Synthesizing Proteins in the Test Tube* (Stanford, CA: Stanford University Press, 1997), 74–83; and Frederic L. Holmes, *Investigative Pathways: Patterns and Stages in the Careers of Experimental Scientists* (New Haven, CT: Yale University Press, 2004), xvi and 93–102.

33. Gilles Deleuze, *Difference and Repetition*, trans. Paul Patton (New York: Columbia University Press, 1994), 22.

34. Ibid., 8.

35. Hermann von Helmholtz, "An Autobiographical Sketch [1891]," trans. Edmund Atkinson, *Science and Culture: Popular and Philosophical Essays*, ed. David Cahan (Chicago: University of Chicago Press,, 1995), 381–92, here 388. See also Holmes, *Investigative Pathways*, xvii.

36. On science and the "patient drudge," see Lorraine Daston and Peter Galison, *Objectivity* (New York: Zone, 2007), 230 and 211–16.

37. On the ontological relation between science and film, see Henri Bergson, *Creative Evolution*, trans. Arthur Mitchell (New York: Holt, 1911), 272–370. Horst Bredekamp's discussion of Galileo's images also refers back to this relation, and Bergson himself, in his discussion of the cinematographical mechanism of thought, already referred to Galileo. It thus seems to be no mere coincidence that the telescope in Bredekamp's study functions as a medium of an aesthetic as well as epistemic acceleration and segmentation. See Horst Bredekamp, *Galilei der Künstler: Der Mond, die Sonne, die Hand*, 2nd ed. (Berlin: Akademie, 2009), esp. 337–40.

38. On the Helmholtz/Proust relation see Anson Rabinbach, *The Human Motor: Energy, Fatigue, and the Origins of Modernity* (Berkeley: University of California Press, 1990), 93–97; Thomas Schestag, "Wiedergefunden: 'du temps perdu,'" *Kritische Beiträge* 4 (1998): 73–94; and Marco Piccolino, "A 'Lost Time' Between Science and Literature: The *'temps perdu'* from Hermann von Helmholtz to Marcel Proust," *Audiological Medicine* 1 (2003): 261–70. On the question of (chrono-) photography in Proust, see Brassaï, *Proust in the Power of Photography*, trans. Richard Howard (Chicago: University of Chicago Press, 2001).

39. See, for example, Charles François-Franck, "Recherches sur les intermittences du pouls, et sur les troubles cardiaques qui les déterminent," *Physiologie Expérimentale: Travaux du laboratoire de M. Marey* 3 (1877): 63–95, here 79 ("intermittence du cœur").

40. On the last point, see Gregor Schiemann, *Hermann von Helmholtz's Mechanism: The Loss of Certainty—A Study on the Transition from Classical to Modern Philosophy of Nature*, trans. Cynthia Klohr (Dordrecht: Springer, 2009), 113–18.

41. On the *entr'acte* and the empty time of the event, see Bernard Groethuysen, "De quelques aspects du temps: Notes pour une phénoménologie du Récit," *Recherches philosophiques* 5 (1935–1936): 139–95, esp. 187–88.

42. Johann Wolfgang Goethe, *Zur Farbenlehre* [1810], ed. Manfred Wenzel (Frankfurt: Deutscher Klassiker Verlag, 1991), 611–13. [The historical part of Goethe's book is not available in English translation.] See also Karl J. Fink, *Goethe's History of Science* (Cambridge: Cambridge University Press, 1991), 72–74.

43. Helmholtz to du Bois-Reymond, March 24, 1852, in *Dokumente einer Freundschaft: Der Briefwechsel zwischen Hermann von Helmholtz und Emil du Bois-Reymond 1846–1894*, ed. Christa Kirsten et al. (Berlin: Akademie-Verlag, 1986), 128.

44. Hermann Helmholtz, "On the Methods of Measuring Very Small Portions of Time, and Their Application to Physiological Purposes," *The London, Edinburgh and Dublin Philosophical Magazine and Journal of Science* 4 (1853): 313–25, here 321.

45. Ibid., 325.

46. Sigfried Giedion, *Mechanization Takes Command: A Contribution to Anonymous History* (Oxford: Oxford University Press, 1948), 20–26; and most recently Hubertus von Amelunxen, Dieter Appelt, and Peter Weibel, eds., *Notation: Kalkül und Form in den Künsten* (Berlin: Akademie der Künste; Karlsruhe: Zentrum für Kunst und Medientechnologie, 2008).

47. Nobert Elias, *Time: An Essay*, trans. Edmund Jephcott (Oxford: Blackwell, 1992), 54.

48. Cornelius Borck, *Hirnströme: Eine Kulturgeschichte der Elektroenzepha-lographie* (Göttingen: Wallstein, 2005); and Stefan Rieger, *Schall und Rauch: Eine Mediengeschichte der Kurve* (Frankfurt: Suhrkamp, 2009). On semiotic and media practices in the sciences, see Timothy Lenoir, ed., *Inscribing Science: Scientific Texts and the Materialities of Communication* (Stanford: Stanford University Press, 1998); Bernhard Siegert, *Passage des Digitalen: Zeichenprak-tiken der neuzeitlichen Wissenschaften 1500–1900* (Berlin: Brinkmann & Bose, 2003); and Michael Franz, Wolfgang Schäffner, Bernhard Siegert, and Robert Stockhammer, eds., *Electric Laokoon: Zeichen und Medien, von der Lochkarte zur Grammatologie* (Berlin: Akademie, 2007).

49. Rheinberger, *Toward a History of Epistemic Things*, 139.

50. Hoff and Geddes, "Graphic Registration before Ludwig"; "Ballistics and the Instrumentation of Physiology: The Velocity of the Projectile and of the Nerve Impulse," *Journal of the History of Medicine and Allied Sciences* 15, no. 2 (1960): 133–46; "The Technological Background of Physiological Discovery: Ballistics and the Graphic Method," *Journal of the History of Medicine and Allied Sciences* 15, no. 4 (1960): 345–63; Wilhelm Blasius, "Die Bestimmung der Leitungsgeschwindigkeit im Nerven durch Hermann v. Helmholtz am Beginn der naturwissenschaftlichen Ära der Neurophysiolo-gie," *Von Boerhaave bis Berger: Die Entwicklung der kontinentalen Physiologie im 18. und 19. Jahrhundert mit besonderer Berücksichtigung der Neurophysiologie*, ed. Karl E. Rothschuh (Stuttgart: G. Fischer, 1964), 71–84; Soraya de Chadarevian, "Graphical Method and Discipline: Self-Recording Instru-ments in Nineteenth-Century Physiology," *Studies in History and Philosophy of Science* 24 (1993): 267–91; Claude Debru, "Helmholtz and the Psychophys-iology of Time," *Science in Context* 14, no. 3 (2001): 471–92; and Stanley Finger and Nicholas J. Wade, "The Neuroscience of Helmholtz and the Theories of Johannes Müller, Part 1: Nerve Cell Structure, Vitalism, and the Nerve Impulse," *Journal of the History of the Neurosciences* 11 (2002): 136–55.

51. Helmholtz to du Bois-Reymond, April 11, 1851, *Dokumente*, 111.

52. Klaus Klauß, "Die erste Mitteilung von H. Helmholtz an die Physika-lische Gesellschaft über die Fortpflanzungsgeschwindigkeit der Reizung in den sensiblen Nerven des Menschen," *NTM* (N.S.) 2 (1994): 89–96; and Henning Schmidgen, "Die Geschwindigkeit von Gedanken und Gefühlen: Die Entwicklung psychophysiologischer Zeitmessungen, 1850–1865," *NTM* (N.S.) 12 (2004): 100–15.

53. On Olga Helmholtz, see Maria Osietzki, "Körpermaschinen und Dampfmaschinen: Vom Wandel der Physiologie und des Körpers unter dem Einfluß von Industrialisierung und Thermodynamik," *Physiologie und industri-elle Gesellschaft*, ed. Philipp Sarasin and Jakob Tanner (Frankfurt: Suhrkamp,

1998), 313–46. On research couples more generally, see Helena Mary Pycior, Nancy G. Slack, and Pnina G. Abir-Am, eds., *Creative Couples in the Sciences* (New Brunswick, NJ: Rutgers University Press, 1996).

54. M. Norton Wise, "What Can Local Circulation Explain? The Case of Helmholtz's Frog-Drawing-Machine in Berlin," *Journal for the History of Science and Technology* 1 (2007): 15–73; and *Neo-Classical Aesthetics of Art and Science: Hermann Helmholtz and the Frog-Drawing Machine* (Uppsala: Uppsala University, 2008).

55. See, however, the fabulous chapter by Peter Berz, "Ein dromologisches Spitzenereignis: der Schuß," *08/15: Ein Standard des 20. Jahrhunderts* (Munich: Fink, 2001), in which Berz brilliantly sketches a media archaeology of this method. In contrast, see Christian Kassung, *Das Pendel: Eine Wissensgeschichte* (Munich: Fink, 2007, 85–149), which tries to develop the perspective of a "transdisciplinary history of knowledge" that proceeds "in exemplary ways, not strictly historical" ones (123).

56. See Volker Aschoff, *Geschichte der Nachrichtentechnik*, vol. 2, *Nachrichtentechnische Entwicklungen in der ersten Hälfte des 19. Jahrhunderts* (Berlin: Springer, 1987); Jakob Messerli, *Gleichmässig, pünktlich, schnell: Zeiteinteilung und Zeitgebrauch in der Schweiz im 19. Jahrhundert* (Zurich: Chronos, 1995); Ian R. Bartky, *Selling the True Time: Nineteenth-Century Timekeeping in America* (Stanford, CA: Stanford University Press, 2000); Peter Galison, *Einstein's Clocks, Poincaré's Maps: Empires of Time* (New York: Norton, 2003); Werner Faulstich, *Medienwandel im Industrie- und Massenzeitalter (1830–1900)* (Göttingen: Vandenhoeck & Ruprecht, 2004); and Ian R. Bartky, *One Time Fits All: The Campaigns for Global Uniformity* (Stanford, CA: Stanford University Press, 2007).

57. Hoff and Geddes, "Graphic Registration before Ludwig," 14–17.

58. Chadarevian, "Graphical Method and Discipline," 273; and Gabriel Finkelstein, "M. du Bois-Reymond Goes to Paris," *The British Journal for the History of Science* 36 (2003): 261–300.

59. Hermann Helmholtz, "Ueber den Stoffverbrauch bei der Muskelaktion," *Archiv für Anatomie, Physiologie und wissenschaftliche Medicin* 12 (1845): 72–83, here 80.

60. See, however, M. Norton Wise, "Mediating Machines," *Science in Context* 2, no. 1 (1988): 77–113; Andrew Pickering, *The Mangle of Practice: Time, Agency, and Science* (Chicago: University of Chicago Press, 1995), esp. 16 and 37–67; Bernhard J. Dotzler, *Papiermaschinen: Versuch über Communication & Control in Literatur und Technik* (Berlin: Akademie, 1996); and Peter Galison, *Image and Logic: The Material Culture of Microphysics* (Chicago: University of Chicago Press, 1997), xvii: "This book is about the machines of physics."

61. Laurence Sterne, *The Life and Opinions of Tristram Shandy, Gentleman,* ed. Melvyn New and Joan New (London: Penguin, 2003), 426.

1. CURVES REGAINED

1. Wassily Kandinsky, *Point and Line to Plane: Contribution to the Analysis of the Pictorial Elements,* trans. Howard Dearstyne and Hilla Rebay (New York: Solomon R. Guggenheim Foundation, 1947), 98.

2. This note was never published in any language other than French; see Hermann Helmholtz, "Deuxième Note sur la vitesse de propagation de l'agent nerveux," *Comptes rendus hebdomadaires des séances de l'Académie des sciences* 33 (1851): 262–65.

3. Hermann Helmholtz, "Note sur la vitesse de propagation de l'agent nerveux dans les nerfs rachidiens," *Comptes rendus hebdomadaires des séances de l'Académie des sciences* 30 (1850): 204–6, here 206.

4. Humboldt to Helmholtz, February 12, 1850, qtd. Leo Königsberger, *Hermann von Helmholtz,* vol. 1 (Braunschweig: Vieweg, 1902), 118. On Humboldt's role in the development of (electro)physiology and the Berlin context see Ilse Jahn, "Die Anfänge der instrumentellen Elektrobiologie in den Briefen Humboldts an Emil Du Bois-Reymond," *Medizinhistorisches Journal* 2 (1967): 135–56.

5. Hermann Helmholtz, "Messungen über den zeitlichen Verlauf der Zuckung animalischer Muskeln und die Fortpflanzungsgeschwindigkeit der Reizung in den Nerven," *Archiv für Anatomie, Physiologie und wissenschaftliche Medicin* (1850): 276–364.

6. Helmholtz, "Deuxième Note," 265.

7. Ibid., 263.

8. Ibid.

9. Etienne-Jules Marey, "Reproduction typographique des photographies, procédé de M. Ch. Petit," *Comptes rendus hebdomadaires des séances de l'Académie des sciences* 95 (1882): 583–85. On this issue, see Maurice Crosland, *Science under Control: The French Academy of Sciences 1795–1914* (Cambridge: Cambridge University Press, 1992), 258. Crosland suggests that it is only after this exception that the *Comptes rendus* accepted illustrations of any kind at all. That is true in the case of photographs. Marey marks this beginning. He presents a new method for reproducing photographs in print. Other kinds of figures, however (diagrams, for example), were reproduced in the *Comptes rendus* as early as the 1870s.

10. Holmes, *Investigative Pathways,* 29–33.

11. Helmholtz, "Deuxième Note," 263.

12. Proceedings of the session of September 1, 1851, Hermann Helmholtz, "Vitesse de propagation du système nerveux [*sic*]," 1, Archives de l'Académie des sciences, Paris.

13. Rudolf Schelske, "Neue Messungen der Fortpflanzungsgeschwindig-keit des Reizes in den menschlichen Nerven," *Archiv für Anatomie, Physiologie und wissenschaftliche Medicin* (1864): 151–73, here 173. Another curve drawing has been preserved in the correspondence between Helmholtz and du Bois-Reymond. This drawing was made specifically for du Bois-Reymond in July 1852. It shows "secondary muscle contractions" and is thus not directly related to Helmholtz' time measurements in nerves. See *Dokumente*, 134. In addition, some curve drawings by Helmholtz are kept in the Turin collection of physiological recordings started by Angelo Mosso in the 1880s, a collection intended to be an important element of the library of Mosso's physiological institute. On this point, see Philipp Felsch, *Laborlandschaften: Physiologische Alpenreisen im 19. Jahrhundert* (Göttingen: Wallstein, 2007), 129 and 99, fig. 19. These "frog curves" were transmitted to Mosso only after Helmholtz's death. They are not dated and, according to Felsch, it is unclear for what purpose they were recorded.

14. Hermann Helmholtz, "Deuxième Note sur la vitesse de propagation de l'agent nerveux. Par M. H. Helmholtz à Königsberg" [Letter to Arago], Helmholtz Papers, NL Helmholtz 526, Archives of the Berlin-Brandenburg Academy of Science, Berlin.

15. It was not until a year later that Helmholtz presented his "new method" to the German-speaking public in a revised, considerably extended and, this time, illustrated journal article. See Helmholtz, "Messungen über Fortpflanzungsgeschwindigkeit der Reizung in den Nerven (Zweite Reihe)," *Archiv für Anatomie, Physiologie und wissenschaftliche Medicin* (1852): 199–216. Some passages of this article are identical to the text of the "Deuxième Note." It is unclear, however, whether those passages were adopted from the "Deuxième Note" or whether there was a common urtext that was eventually lost.

16. Hermann Helmholtz, "Messungen über den zeitlichen Verlauf der Zuckung animalischer Muskeln und die Fortpflanzungsgeschwindigkeit der Reizung in den Nerven," *Archiv für Anatomie, Physiologie und wissenschaftliche Medicin* (1850): 276–364.

17. On this point, see also Marco Piccolino, "A 'Lost Time' Between Science and Literature: The *'temps perdu'* from Hermann von Helmholtz to Marcel Proust," *Audiological Medicine* 1 (2003): 261–70.

18. Helmholtz, "Deuxième Note," 263.

19. Ibid.

20. Ibid.

21. Ibid.

22. Nicholas Jardine, "The Laboratory Revolution in Medicine as Rhetori-cal and Aesthetic Accomplishment," in *The Laboratory Revolution in Medicine*, ed.

Andrew Cunningham and William Perry (Cambridge: Cambridge University Press, 1992), 304–23. The literature on visualization in the history of the life sciences is quite extensive. On the exemplary case of embryology, see Nick Hopwood, " 'Giving Body' to Embryos: Modeling, Mechanism, and the Microtome in Late Nineteenth-Century Anatomy," *Isis* 90 (1999): 463–96. On Darwinism see Howard E. Gruber, "Darwin's 'Tree of Nature' and Other Images of Wide Scope," *On Aesthetics in Nature*, ed. Judith Wechsler (Cambridge: MIT Press, 1978), 121–40; and Phillip Prodger, "Illustration as Strategy in Charles Darwin's 'Expression of the Emotions in Man and Animals,' " *Inscribing Science*, ed. Timothy Lenoir, 139–81. In addition, see Horst Bredekamp, *Darwins Korallen: Frühe Evolutionsmodelle und die Tradition der Naturgeschichte* (Berlin: Wagenbach, 2005); Jonathan Smith, *Charles Darwin and Victorian Visual Culture* (Cambridge: Cambridge University Press, 2006); and Julia Voss, *Darwin's Pictures: Views of Evolutionary Theory, 1837–1874*, trans. Lori Lantz (New Haven, CT: Yale University Press, 2010). See also Fae Brauer and Barbara Larson, eds., *The Art of Evolution: Darwin, Darwinisms, and Visual Cultures* (Hanover, NH: University Press of New England, 2009); and Pamela Kort and Max Hollein, eds., *Darwin: Art and the Search for Origins* (Cologne: Wienand, 2009).

23. Johann Nepomuk Czermak, "Die Physiologie als allgemeines Bildungs-Element," 1870, *Gesammelte Schriften* (Leipzig: Engelmann, 1879), vol. 2, 105–18, here 116.

24. Emil du Bois-Reymond, "Der physiologische Unterricht sonst und jetzt," *Reden von Emil du Bois-Reymond in zwei Bänden*, ed. Estelle du Bois-Reymond (Leipzig: Veit, 1912), vol. 1, 630–53, here 651.

25. Du Bois-Reymond, "Der physiologische Unterricht sonst und jetzt," 650. On the architectures and techniques of visual instruction in nineteenth-century experimental physiology, see Henning Schmidgen, "Pictures, Preparations, and Living Processes: The Production of Immediate Visual Perception (*Anschauung*) in Late-Nineteenth-Century Physiology," *Journal of the History of Biology* 37 (2004): 477–513.

26. Helmholtz to du Bois-Reymond, June 20, 1852, *Dokumente*, 131.

27. Helmholtz, "Deuxième Note," 263–64.

28. "Myograph" literally means "muscle writer." According to the *Oxford English Dictionary*, the term refers generally to an "an instrument for displaying or recording muscular contractions and relaxations." The *OED* considers the term a derivation from the French word *myographe*, used in the 1820s to designate "someone who describes muscles," and credits Etienne-Jules Marey with introducing it for denoting an instrument in "1866 or earlier." In the German context, Helmholtz introduced the term "myographion" in

the mid-1850s, that is, shortly after the period I focus on in this book. In a letter to du Bois-Reymond, dated June 13, 1854, Helmholtz writes: "The physiological institute in Gießen has constructed their own frog drawing apparatus or 'myographion,' as I would like to pompously call it from now on" (*Dokumente*, 144).

29. Etienne-Jules Marey, *Du mouvement dans les fonctions de la vie* (Paris: Ballière, 1868), 224.

30. Helmholtz, "Messungen," 206.

31. Thomas Young, *A Course of Lectures on Natural Philosophy and the Mechanical Arts* (London: Johnson, 1807), vol. 1, 772, plate 15, fig. 198.

32. See Stanley Joel Reiser, *Medicine and the Reign of Technology* (Cambridge: Cambridge University Press, 1979), 100.

33. Helmholtz, "Deuxième Note," 264.

34. Ibid., 265.

35. Ibid.

36. Ibid.

37. Ibid.

2. SEMIOTIC THINGS

1. Gilles Deleuze and Félix Guattari, *Anti-Oedipus: Capitalism and Schizophrenia* 1, trans. Robert Hurley, Mark Seem, and Helen R. Lane (Minneapolis: University of Minnesota Press, 1983), 39.

2. Bruno Latour, "Circulating Reference: Sampling the Soil in the Amazonas Forest," *Pandora's Hope: Essays on the Reality of Science Studies* (Cambridge: Harvard University Press, 1999), 24–79, here 32.

3. Proceedings of the session of September 1, 1851, Hermann Helmholtz, "Vitesse de propagation du système nerveux [*sic*]," 3, Archives de l'Académie des sciences, Paris.

4. Bruno Latour, "Drawing Things Together," Michael Lynch and Steve Woolgar, eds., *Representation in Scientific Practice* (Cambridge: MIT Press, 1990), 19–68.

5. Johann Joseph von Prechtl, "Folien," *Technologische Encyclopädie, oder alphabetisches Handbuch der Technologie, der technischen Chemie und des Maschinenwesens*, vol. 6 (Stuttgart: Cotta, 1835), 261–64, here 264; and Anonymous, "Aufpausen," *Meyers Großes Konversations-Lexikon*, vol. 2, 6th ed. (Leipzig: Bibliographisches Institut, 1905), 96.

6. Helmholtz to du Bois-Reymond, June 12, 1851, *Dokumente*, 115.

7. Helmholtz, "Vitesse de propagation du système nerveux," 3.

8. "Filmstrip" (*Filmstreifen*) is the telling name the Staatsbibliothek in Berlin uses in the finding aid for the correspondence between Helmholtz and

du Bois-Reymond for describing the curve drawings that have survived in this collection. Bernhard Siegert has drawn my attention to the similarity between Helmholtz's fish glue and the isinglass described by Herman Melville, in a passage of *Moby Dick*, as that "infinitely thin, transparent substance" that can be scraped off from whales' dead bodies. Similar to fish glue, Melville's isinglass can be dried and preserved as leaves. For the narrator of *Moby Dick*, these leaves function as bookmarks in his whale books. Placed on the printed page they exert "a magnifying influence" that facilitates reading. In its original form, however, isinglass shows and protects "mysterious ciphers" on the body of the whale that partly resemble engravings, partly "pyramids hieroglyphics." See Herman Melville, *Moby Dick, or, The Whale* (London: Penguin, 2003), 332–33. On the whale, see also below, chapter 6.

9. Georges Didi-Huberman, ed., *L'empreinte* (Paris: Ed. du Centre Georges Pompidou, 1997). With respect to science and medicine, see Marie-Dominique de Teneuille and Quentin Bajac, eds., *À fleur de peau: Le moulage sur nature au XIXe siècle* (Paris: Réunion des musées nationaux, 2001); and, more generally, Sybille Krämer, Werner Kogge, and Gernot Grube, eds., *Spur: Spurenlesen als Orientierungstechnik und Wissenskunst* (Frankfurt: Suhrkamp, 2007).

10. Helmholtz, "Messungen (Zweite Reihe)," 215.

11. Angelo Mosso, *L'Institut Physiologique de l'Université de Turin* (Turin: Bona, 1894), 27. See also Heinz Schröder, *Carl Ludwig: Begründer der messenden Experimentalphysiologie, 1816–1895* (Stuttgart: Wissenschaftliche Verlagsgesellschaft, 1967), 108 and fig. 14, as well as Felsch, *Laborlandschaften*, 13.

12. In the 1870s and 1880s, du Bois-Reymond used the expression "autographical method" to refer to the use of kymographs and myographs in physiological research; cf., for example, his "Gedächtnisrede auf Hermann von Helmholtz: Gehalten in der Leibniz-Sitzung der Akademie der Wissenschaften am 4. Juli 1895," *Reden* 2:516–570, here 529.

13. Helmholtz, "Deuxième Note," 264.

14. Emil du Bois-Reymond, *Untersuchungen über thierische Elektricität* (Berlin: Reimer, 1848–1884), vol. 1, xxix.

15. Ibid., xxxiii and xxvi.

16. Adolphe Quételet, *A Treatise on Man and the Development of his Faculties*, trans. Robert Knox (Edinburgh: Chambers, 1842), 61.

17. On Quételet and the "method of curves" (du Bois-Reymond does not use the term until 1884), see du Bois-Reymond, *Untersuchungen*, vol. 2/2, 505. Quételet, for his part, speaks of "lines." On this issue, see also Chadarevian, "Graphical Method and Discipline," 284, footnote 40.

18. Du Bois-Reymond, *Untersuchungen*, vol. 1, xxvii.

19. Proceedings of the session of September 1st, 1851, Hermann Helmholtz, "Explication des épreuves," 1, Archives de l'Académie des sciences, Paris.

20. Ibid.

21. Ibid.

22. Helmholtz to du Bois-Reymond, September 17, 1850, *Dokumente,* 106.

23. Helmholtz to O. Helmholtz, August 6, 1851, in *Letters of Hermann von Helmholtz to his Wife, 1847–1859,* ed. Richard L. Kremer (Stuttgart: Steiner, 1990), 52.

24. Helmholtz to du Bois-Reymond, April 11, 1851, *Dokumente,* 111.

25. Galison, *Image and Logic,* 74–75.

26. See Holmes, *Investigative Pathways,* 56–71.

3. A RESEARCH MACHINE

1. "L'écart est une opération." Marcel Duchamp, *Le surrealisme au service de la Révolution,* no. 5 (1933), 1.

2. Gilles Deleuze, "What is a dispositif?" *Michel Foucault: Philosopher,* ed. and trans. Timothy J. Armstrong (New York: Routledge, 1992), 159–68, here 159.

3. Hermann Helmholtz, *De fabrica systematis nervosi evertebratorum* (Berlin: Nietack, 1842).

4. Hermann Helmholtz, "Ueber das Wesen der Fäulnis und Gärung," *Archiv für Anatomie, Physiologie und wissenschaftliche Medicin* 10 (1843): 453–62; "Ueber den Stoffverbrauch bei der Muskelaktion," *Archiv für Anatomie, Physiologie und wissenschaftliche Medicin* 12 (1845): 72–83; "Waerme, physiologisch," *Encyclopaedisches Woerterbuch der medicinischen Wissenschaften* (Berlin: Veit & Co., 1846) 523–67; *Über die Erhaltung der Kraft: Eine physikalische Abhandlung* (Berlin: Reimer, 1847); "Ueber die Wärmeentwickelung bei der Muskelaction," *Archiv für Anatomie, Physiologie und wissenschaftliche Medicin* 15 (1848), 144–64.

5. Important accounts of these investigations can be found in Timothy Lenoir, *The Strategy of Life: Teleology and Mechanics in Nineteenth-Century German Biology* (Chicago: University of Chicago Press, 1989), 197–245; and Richard L. Kremer, *The Thermodynamics of Life and Experimental Physiology, 1770–1880* (New York: Garland, 1990), 237–55 and 292–307. See, however, the much earlier text, John F. Fulton, *Muscular Contraction and the Reflex Control of Movement* (London: Williams & Wilkins, 1926), 42–47; and Everett Mendelsohn (*Heat and Life: The Development of the Theory of Animal Heat*

[Cambridge: Harvard University Press, 1964]), who presents Helmholtz's studies of muscle action as putting an end to all traditional theories of organic heat.

6. Hermann Helmholtz, "Vorläufiger Bericht über die Fortpflanzungsgeschwindigkeit der Nervenreizung," *Archiv für Anatomie, Physiologie und wissenschaftliche Medizin* 17 (1850): 71–73, here 71.

7. The machine is described in chapter 2 of *Locus Solus*, trans. Rupert Copeland Cuningham (Richmond: Oneworld Classics, 2008).

8. See the 1963 French edition published by Gallimard.

9. Hermann Helmholtz to Olga von Velten, July 18, 1847, *Letters to his Wife*, 43.

10. One of the earliest uses of the term "graphical method" (*graphische Methode*) in the German-speaking world can be found in Rudolf Schelske, "Neue Messungen der Fortpflanzungsgeschwindigkeit des Reizes in den menschlichen Nerven," *Archiv für Anatomie, Physiologie und wissenschaftliche Medicin* (1864): 151–73, here 153. Carl Ludwig, in his pioneering article of 1847, had only spoken of "graphical representation" (*graphische Darstellung*) and was not quite explicit on whether this referred to the curve recording on the kymograph drum or to the reproduction of the recorded curve in printed form. Some years later, Karl Vierordt, in his book on the sphygmograph, presents this instrument as a contribution to a new "method of visual representation" (*Die Lehre vom Arterienpuls in gesunden und kranken Zuständen* [Braunschweig: Vieweg, 1855]). Shortly after Schelske, the term *graphische Methode* can also be found in Joseph Pisko, *Die neueren Apparate der Akustik: Für Freunde der Naturwissenschaft und der Tonkunst* (Vienna: Gerold's Sohn, 1865), 4. In the French context, however, the term "méthode graphique" appears much earlier. Antoine Jourdan, for example, uses the expression to denote chemical notations for compound substances (*Dictionnaire raisonné, etymologique, synonymique et polyglotte, des termes usités dans les sciences naturelles*, vol. 1 [Paris: Baillière, 1834], 517).

11. Helmholtz to O. Helmholtz, July 18, 1848, *Letters to his Wife*, 43.

12. Helmholtz to du Bois-Reymond, October 14, 1849, *Dokumente*, 87.

13. Helmholtz, "Vorläufiger Bericht," 72.

14. Helmholtz, "Messungen," 286.

15. Ibid., 287.

16. Ibid.

17. On this issue, see Kremer, *The Thermodynamics of Life*, 308–33 and 461–63 (figs. 6–8). In 1836, Eduard Weber and Wilhelm Weber published their anatomical-physiological investigation of the mechanics of the human gait, *Mechanik der menschlichen Gehwerkzeuge: Eine anatomisch-physiologische Untersuchung* (Göttingen: Dieterich, 1836).

18. Helmholtz, "Messungen," 277.

19. Ibid.

20. Concerning the complex relation between force, energy, and work in Helmholtz's time experiments see chapter 8, " 'A Spectacle for the Gods,' " of M. Norton Wise, *Bourgeois Berlin and Laboratory Science* (in preparation).

21. Eduard Weber, "Muskelbewegung," *Handwörterbuch der Physiologie mit Rücksicht auf physiologische Pathologie*, vol. 3/2, ed. Rudolph Wagner (Braunschweig: Vieweg, 1846), 1–122, here 2.

22. Weber, "Muskelbewegung," 3.

23. Helmholtz, "Messungen," 290.

24. Pouillet, "Note sur un moyen de mesurer des intervalles de temps extrémement courts."

25. Thomas Willis, "Of Musculary Motion," *Five Treatises* (London: Clavell, 1681), 34–48, here 40. Georges Canguilhem quotes this passage in his *La formation du concept de réflexe aux XVIIe et XVIII siècles*, 2nd ed. (Paris: Vrin, 1977), 178–79.

26. Helmholtz, "Messungen," 278.

27. Ibid., 289.

28. Ibid., 290.

29. Carl Friedrich Gauß and Wilhelm Weber, "Results of the Observations Made by the Magnetic Association in the Year 1836," trans. W. Francis with Prof. Lloyd and Maj. Sabine, *Scientific Memoirs, Selected from the Transactions of Foreign Academies of Science and Learned Societies, and from Foreign Journals*, ed. Richard Taylor, vol. 2 (1841), 20–97, here 24; Johann Christian Poggendorff, "Ein Vorschlag zum Messen der magnetischen Abweichung," *Annalen der Physik und Chemie* 7 (1826): 121–30.

30. Gauß and Weber, "Results 1836," 26. On standardization in this context, see Matthias Dörries, "La standardisation de la balance de torsion dans les projets européens sur le magnétisme terrestre," in *Restaging Coulomb: Usages, controverses et réplications autour de la balance de torsion*, ed. Christine Blondel and Matthias Dörries (Florence: Olschki, 1994), 121–49. On the cooperation between Gauß and Göttingen-based instruments makers, see Klaus Hentschel, *Gaußens unsichtbare Hand: Der Universitäts-Mechanicus und Maschinen-Inspector Moritz Meyerstein* (Göttingen: Vandenhoeck & Ruprecht, 2005).

31. Poggendorff, "Vorschlag," 122–23, as well as Gauß and Weber, "Results 1836," 32.

32. Carl Friedrich Gauß, [Announcement: Intensitas vis magneticae terrestris ad mensuram absolutam revocata, 1832 Dec.], *Werke*, vol. 5 (Göttingen: Dieterich, 1867), 293–304, here 301.

33. Wilhem Weber, "Ueber die Elasticität der Seidenfäden," *Annalen der Physik und Chemie* 110/2 (1835): 247–57, as well as Gauß and Weber, "Results

1836," 50. [The passage is only partially translated, cf. the original German, *Resultate aus den Beobachtungen des magnetischen Vereins im Jahre 1836* (Göttingen: Dieterich, 1837), 45.]

34. Carl Friedrich Gauß, "Erdmagnetismus und Magnetometer," 1836, *Werke*, vol. 5, 315–44, here 338. On synchronizing the observations made in the *Magnetische Verein*, see Matthias Dörries, "Balances, Spectroscopes, and the Reflexive Nature of Experiment," *Studies in History and Philosophy of Science* 25, no. 1 (1994): 1–36, esp. 15–21.

35. Carl Friedrich Gauß and Wilhelm Weber, *Resultate aus den Beobachtungen des magnetischen Vereins im Jahre 1837* (Göttingen: Dieterich, 1838), 17–18.

36. Aschoff, *Geschichte der Nachrichtentechnik*, vol. 2, 65–129. For the principal sources concerning the history of the telegraph constructed by Gauß and Weber, see the appendix in Ernst Feyerabend, *Der Telegraph von Gauß und Weber im Werden der elektrischen Telegraphie* (Berlin: Reichspostministerium, 1933).

37. On this issue, see Gauß, [Announcement: Intensitas], 302.

38. See Diane Greco Josefowicz, "Experience, Pedagogy, and the Study of Terrestrial Magnetism," *Perspectives on Science* 13, no. 4 (2006): 452–94.

39. Hermann Helmholtz, "Ueber die Methoden, kleinste Zeittheile zu messen, und ihre Anwendung für physiologische Zwecke," *Koenigsberger naturwissenschaftliche Unterhaltungen* 2 (1850): 169–89, here 177. This passage is missing from the English translation of Helmholtz's text, "On the Methods of Measuring Very Small Portions of Time, and Their Application to Physiological Purposes," *The London, Edinburgh and Dublin Philosophical Magazine and Journal of Science* 4 (1853): 313–25, where it would figure on page 318.

40. See Alexandre Métraux, "Le régime scopique du XIXe siècle et la vision selon Hermann von Helmholtz," *Philosophiae Scientiae* 7, no. 1 (2003): 151–66.

41. Marcel Proust, Letter to Camille Vettard, March 1922, *Letters of Marcel Proust*, trans. and ed. Mina Curtiss (New York: Vintage, 1966), 405, and *In Search of Lost Time*, vol. 6, *Finding Time Again*, trans. Ian Patterson (London: Penguin, 2003), 342.

42. On this point, see Henning Schmidgen, *Hirn und Zeit*.

43. Wilhelm Weber, "Beschreibung eines kleinen Apparats zur Messung des Erdmagnetismus nach absolutem Maass für Reisende," *Resultate aus den Beobachtungen des magnetischen Vereins im Jahre 1836*, ed. Carl Friedrich Gauß and Wilhelm Weber (Göttingen: Diederichs, 1837), 63–89; and Carl Friedrich Gauß, "Anleitung zur Bestimmung der Schwingungsdauer einer Magnet-

nadel," *Resultate aus den Beobachtungen des magnetischen Vereins im Jahre 1837*, ed. Carl Friedrich Gauß and Wilhelm Weber (Göttingen: Diederichs, 1838), 58–80.

44. Helmholtz, "Messungen," 300.

45. Ibid., 298.

46. Ibid., 339.

47. Ibid., 351.

48. Ibid., 279–80.

49. Ibid., 280.

50. Ibid., 281.

51. Ibid., 284.

52. Ibid., 281.

53. Ibid., 283.

54. Ibid., 280.

55. Ibid., 283.

56. Ibid.

57. Oskar Langendorff, *Physiologische Graphik: Ein Leitfaden der in der Physiologie gebräuchlichen Registrirmethoden* (Leipzig: Deuticke, 1891), 10.

58. Helmholtz, "Messungen," 283–84.

59. Ibid., 284.

60. Ibid., 277.

61. Helmholtz to du Bois-Reymond, October 14, 1849, *Dokumente*, 87.

62. Ibid., 88.

63. Ibid.

64. Helmholtz, "Messungen," 332–33.

65. Helmholtz to du Bois-Reymond, January 15, 1850, *Dokumente*, 91.

66. Helmholtz to du Bois-Reymond, October 14, 1849, ibid., 88.

67. Helmholtz to du Bois-Reymond, January 15, 1850, ibid., 91.

68. On the notion of the "bachelor machine," see Michel Carrouges, *Les machines célibataires* (Paris: Arcanes, 1954); and Harald Szeemann, ed., *Le machine celibi/ The Bachelor Machines* (Venice: Alfieri, 1975).

69. Henning Schmidgen, "The Donders Machine: Matter, Signs, and Time in a Physiological Experiment, c. 1865," *Configurations* 13, no. 2 (2005/2007): 211–56.

70. Helmholtz, "Messungen," 337.

71. Hermann Helmholtz and Olga Helmholtz, Laboratory logbook, March 29 to June 9, 1850, *Archiv der Berlin-Brandenburgischen Akademie der Wissenschaften*, Helmholtz Papers, NL Helmholtz 547: 13, 14, and 46. This laboratory logbook is already mentioned by Königsberger (*Hermann von Helmholtz*, vol. 1, 113) who, however, attributes it to Olga Helmholtz alone.

72. H. and O. Helmholtz, Laboratory logbook 8 and 33.

73. Helmholtz, "Messungen," 345–50.

74. Ibid., 334.

75. Ibid.

76. Ibid., 358.

77. Ibid., 344.

78. Ibid., 358.

79. Ibid., 364.

80. Helmholtz, "Wärmeentwickelung bei der Muskelaction," 164.

81. Henri Raczymow, *Le Paris littéraire et intime de Marcel Proust* (Paris: Parigramme, 1997), 41.

82. Marcel Proust, "Days of Reading (I)," *Against Sainte Beuve and Other Esssays*, trans. John Sturrock (London: Penguin, 1988), 194–226, here 202.

83. Ibid.

84. Marcel Proust, "The Prisoner," *In Search of Lost Time*, vol. 5, trans. Carol Clark and Peter Collier (London: Penguin, 2002), 1–384, here 18.

4. NETWORKS OF TIME, NETWORKS OF KNOWLEDGE

1. Air, "Electronic Performers," track 1 on the album *10000 Hz Legend* (Virgin Records, 2001).

2. Ulrich Päßler, *Ein "Diplomat aus den Wäldern des Orinoko": Alexander von Humboldt als Mittler zwischen Preußen und Frankreich* (Stuttgart: Steiner, 2009). On the *Physikalische Gesellschaft*, see Horst Kant, *Ein "mächtig anregender Kreis":Die Anfänge der Physikalischen Gesellschaft zu Berlin* (Berlin: Max-Planck-Institut für Wissenschaftsgeschichte, 2002).

3. Du Bois-Reymond to Helmholtz, March 19, 1850, *Dokumente*, 93.

4. Du Bois-Reymond to Helmholtz, March 18, 1851, *Dokumente*, 106.

5. See Helmholtz to A. F. J. Helmholtz, March 19, 1850: "The apparatuses that I have needed for my work so far have been manufactured pretty well here, too" (qtd. Leo Königsberger, *Hermann von Helmholtz*, vol. 1, 121).

6. For a general presentation, see Sven Dierig, *Wissenschaft in der Maschinenstadt: Emil Du Bois-Reymond und seine Laboratorien in Berlin* (Göttingen: Wallstein, 2006).

7. On Veit, see Anne-Katrin Ziesak, *Walter de Gruyter, Publisher, 1749–1999*, trans. Rhodes Barrett (Berlin and New York: de Gruyter, 1999), 103–38.

8. Pouillet, "Note sur un moyen de mesurer des intervalles de temps extrémement courts."

9. Louis Breguet, "Note sur un appareil destiné à mesurer la vitesse d'un projectile dans différents points de sa trajectoire," *Comptes rendus hebdomadaires des séances de l'Académie des sciences de Paris* 20 (1845): 157–62.

10. Ibid., 157.

11. Charles Wheatstone, "Note on the Electro-magnetic Chronoscope," *Electrical Magazine* 2 (1845): 86–93, here 87. Cf. *Comptes rendus hebdomadaires des séances de l'Académie des sciences* 20 (1845): 1554–61.

12. Ibid., 92–93.

13. Regarding the concept of the "lineage" of technological objects, see Gilbert Simondon, *Du mode d'existence des objets techniques,* 3rd ed. (Paris: Aubier, 1989), 40–49. On the history of time-measuring devices in nineteenth-century physiology and psychology, see Horst Gundlach, "Time-Measuring Apparatus in Psychology," *A Pictorial History of Psychology,* ed. Wolfgang G. Bringmann et al. (Chicago: Quintessence, 1997), 111–16; and Rand B. Evans, "Chronoscope," *Instruments of Science,* ed. Robert Bud and Deborah J. Warner (New York: The Science Museum, London and The National Museum of American History, Smithsonian Institution, 1998), 115–16.

14. On the history of electric clocks, see Charles Kenneth Aked, *Electrifying Time: An Exhibition Held at the Science Museum to Commemorate the Centenary of the Death of Alexander Bain, 2nd January, 1877* (Ticehurst: The Antiquarian Horological Society, 1976); *A Conspectus of Electrical Timekeeping* (Ticehurst: The Antiquarian Horological Society, 1976); Willem D. Hackmann, ed., *Alexander Bain's Short History of the Electric Clock* (London: Turner & Devereux, 1973); Frank Hope-Johns, *Electrical Timekeeping* (London: N.A.G. Press, 1949); and J. D. Weaver, *Electrical and Electronic Clocks and Watches* (London: Newnes Technical Books, 1982).

15. Werner Siemens, "Ueber die Anwendung des elektrischen Funkens zu Geschwindigkeitsmessungen," *Annalen der Physik und Chemie* 142, no. 11 (1845): 435–45.

16. Ibid., 436.

17. Artur Fürst, *Werner von Siemens: Der Begründer der modernen Elektrotechnik* (Stuttgart and Berlin: Deutsche Verlagsanstalt, 1916), 39. More concretely, the issue was to test a needle telegraph by Wheatstone in which sender and receiver had to run synchronically. Apparently it was only thanks to Siemens that this requirement was met. See Siegfried von Weiher, *Werner von Siemens: Ein Leben für Wissenschaft, Technik und Wirtschaft* (Göttingen: Musterschmidt, 1970), 20–21.

18. Siemens, "Ueber die Anwendung des elektrischen Funkens," 439.

19. Ibid., 440.

20. Werner Siemens, "Ueber Geschwindigkeitsmessung," *Fortschritte der Physik im Jahre 1845* (Berlin: Reimer, 1847), 47–72.

21. Hermann Helmholtz, "On the Methods of Measuring Very Small Portions of Time, and Their Application to Physiological Purposes," *The London, Edinburgh and Dublin Philosophical Magazine and Journal of Science* 4 (1853): 313–25, here 316.

22. Johannes Müller, *Elements of Physiology*, trans. William Baly (Philadelphia: Lea and Blanchard, 1843), 532.

23. Ibid.

24. Gustav Karsten, "Vorbericht," *Fortschritte der Physik im Jahre 1846* (Berlin: Reimer, 1848), 3–18, here 15.

25. Anonymous, "Progrès des sciences physiques hors de France," *Revue scientifique et industrielle* 11 (1846): 81–96, here 82.

26. Ibid.

27. On Matteucci, see Edwin Clarke and Leon S. Jacyna, *Nineteenth-Century Origins of Neuroscientific Concepts* (Berkeley: University of California Press, 1987), 196–211. On Matteucci and the graphic method, see Hebbel E. Hoff and Leslie A. Geddes, "Graphic Registration Before Ludwig: The Antecedents of the Kymograph," *Isis* 50, no. 159 (1959): 5–21, here 14–17. Richard Kremer (*The Thermodynamics of Life*, 328, n. 47) suggests that Matteucci used the recording instrument in question even earlier: "Carlo Matteucci in 1844 had invented a recording myograph;" for a contrasting view, see Schröder, *Carl Ludwig*, 104–14.

28. Carlo Matteucci, "Electro-Physiological Researches, Seventh and Last Series Upon the Relation between the Intensity of the Electric Current, and That of the Corresponding Physiological Effect," *Philosophical Transactions of the Royal Society London* 137 (1847): 243–48, here 246.

29. Ibid.

30. Ibid.

31. Ibid.

32. Ibid.

33. Ludwig would refer to Watt only later in this textbook, *Lehrbuch der Physiologie des Menschen* (Heidelberg: Winter, 1852–56), vol. 1, 333 (with reference to Helmholtz), and vol. 2, 85 (with reference to his own investigations).

34. Matteucci, "Electro-Physiological Researches," 246.

35. Ibid., 247.

36. Ibid.

37. Du Bois-Reymond, *Untersuchungen*, vol. 1, 5.

38. On this topic, see Finkelstein, "M. du Bois-Reymond Goes to Paris," esp. 264–69 and 292–94.

39. Carlo Matteucci, "Nouvelles recherches sur l'électricité animale: Du courant musculaire et du courant propre: Extrait d'une lettre de M. Mateucci

à M. de Humboldt (Pisa, 27.3.1845)," *Comptes rendus hebdomadaires des séances de l'Académie des sciences* 20 (1845): 1096–98.

40. Du Bois-Reymond to Humboldt, May 20, 1845 (unsent letter), *Briefwechsel zwischen Alexander von Humboldt und Emil du Bois-Reymond*, ed. Ingo Schwarz and Klaus Wenig (Berlin: Akademie, 1997), 77.

41. Müller, *Elements of Physiology*, 549. See also the account given by du Bois-Reymond in his *Gedächtnisrede auf Johannes Müller* (Berlin: Dümmler, 1860).

42. Carlo Matteucci, *Traité des phénomènes électro-physiologiques des animaux, suivi d'Études anatomiques sur le système nerveux et sur l'organe électrique de la torpille* (Paris: Fortin, Masson & Ciem 1844), xvii.

43. Du Bois-Reymond to Helmholtz, March 19, 1850, *Dokumente*, 93. On the experimental aesthetics of du Bois-Reymond, see Brigitte Lohff, "Emil du Bois-Reymond's Theorie des Experiments," *Naturwissen und Erkenntnis im 19. Jahrhundert: Emil du Bois-Reymond*, ed. Gunter Mann (Hildesheim: Gerstenberg, 1981), 117–28.

44. Helmholtz to du Bois-Reymond, April 11, 1851, ibid., 111.

45. Du Bois-Reymond to Humboldt, May 20, 1845 (unsent letter), 77.

46. Ibid.

47. This is one of the roots of du Bois-Reymond's later contributions to the history of science. On this topic, see Canguilhem, *La formation du concept de réflexe*, 138–43 and 158–59; Dietrich von Engelhardt, *Historisches Bewußtsein in der Naturwissenschaft von der Aufklärung bis zum Positivismus* (Freiburg and Munich: Alber, 1979); and Nicholas Jardine, "The Mantle of Müller and the Ghost of Goethe: Interactions between the Sciences and Their Histories," Donald R. Kelley, ed., *History and the Disciplines. The Reclassification of Knowledge in Early Modern Europe* (Rochester: University of Rochester Press, 1997), 297–17.

48. Du Bois-Reymond to Humboldt, May 20, 1845 (unsent letter), 77.

49. Gustav Karsten, "Vorbericht," *Fortschritte der Physik im Jahre 1845* (Berlin: Reimer, 1847), 3–10, here 4.

50. Ibid.

51. Ibid.

52. Ibid., 5.

53. Emil du Bois-Reymond, "Elektrophysiologie," *Die Fortschritte der Physik im Jahre 1847*, (Berlin: Reimer, 1850), 392–450, here 402–3.

54. Ibid., 435 and 426.

55. Humboldt to du Bois-Reymond, May 8, 1849, *Briefwechsel zwischen Alexander von Humboldt und Emil du Bois-Reymond*, 86 and 88.

56. See *Dokumente*, 105, 133, and 211.

57. Hermann Helmholtz, "Die Resultate der neueren Forschungen über thierische Electricität," *Allgemeine Monatsschrift für Wissenschaft und Literatur* (1852): 294–309, here 297–98. More generally, see *Dokumente*, 105, 133, and 211.

58. Helmholtz to du Bois-Reymond, July 21, 1847, *Dokumente*, 82.

59. Wilhelm Oelschläger, "Das Hipp'sche Chronoskop zur Messung der Fallzeit eines Körpers und zu Versuchen über die Geschwindigkeit der Flintenkugeln, etc.," *Polytechnisches Journal* 14, no. 114 (1849): 255–59.

60. Wilhelm Wundt, *Grundzüge der physiologischen Psychologie* (Leipzig: Engelmann, 1874), 772.

61. See *Prix-Courant de la Fabrique de Télégraphes & Appareils électriques: Fondée par M. Hipp, en 1860: Catalogue E: Horloges électriques* (Neuchâtel: Peyer, Favarger & Compagnie, 1902), 31.

62. Anonymous, "Der neue Buchstaben-Schreibtelegraph des Mechanikus Mathias Hipp aus Reutlingen," *Illustrirte Zeitung* 17, no. 442 (1851): 491–92, here 491.

63. Ibid., 492. On the history of the copying telegraph, see also Christian Kassung, "Isochronie und Synchronie: Zur apparativen und epistemologischen Genese des Kopiertelegraphen," *Lebendige Zeit: Wissenskulturen im Werden*, ed. Henning Schmidgen (Berlin: Kadmos 2005), 196–209.

64. Anonymous, "Der neue Buchstaben-Schreibtelegraph," 492.

65. With respect to Hipp's role in the Swiss context, see Generaldirektion PTT, ed., *Hundert Jahre elektrisches Nachrichtenwesen in der Schweiz, 1852–1952*, vol. 1: *Telegraph* (Bern: Generaldirektion PTT, 1952), 152–79.

66. On the history of the Neuchâtel observatory, see *L'observatoire cantonal neuchâtelois, 1858–1912: Souvenir de son Cinquantenaire et de l'Inauguration du Pavillon Hirsch* (Neuchâtel: Attinger, 1912); and Edmond Guyot, "L'observatoire cantonal de Neuchâtel, 1858–1938: Son histoire, son organisation et ses buts actuels," *Bulletin de la Société neuchâteloise des Sciences naturelles* 63 (1938): 5–36.

67. Messerli, *Gleichmässig, pünktlich, schnell*, 74.

68. Anonymous, *Notice historique concernant les horloges électriques du système inventé et adopté par M. Hipp*, 1881. (Neuchâtel: Peyer, Favarger & Compagnie, n.d.). See also Hipp's extended series of articles on electric clocks, "Die Electricität als Motor für Uhren," *Deutsche Uhrmacher-Zeitung* 3 (1879) and 4 (1880).

69. Alfred Niaudet, "L'unification de l'heure à Paris," *La Nature* 9 (1881): 99–102. On the same issue, see David Aubin, "The Fading Star of the Paris Observatory in the Nineteenth Century: Astronomers' Urban Culture of Circulation and Observation," *Osiris* 18 (2003): 79–100.

70. On the pneumatic clocks at the World's Fair, see Alphonse F. Noguès, "Les horloges pneumatiques à l'Exposition universelle," *La Nature* 6 (1878): 161–62. On the pneumatic clock system that distributed time in the French capital, see Edouard Hospitalier, "Les horloges pneumatiques: La distribution de l'heure à domicile," *La Nature* 8 (1880): 407–9; and Anonymous, "Das Betriebssystem der pneumatischen Uhren in Paris," *Deutsche Uhrmacherzeitung* 6 (1882): 24–25, 31–32, and 39–40.

71. Eugène Bourdon, "Horloge à tube flexible et son moteur hydro-pneumatique," *Revue chronométrique* 24 (1879): 202–7; and Anonymous, "Horloge pneumatique de M. E.-J. Muybridge de San Francisco," *Revue chronométrique* 24 (1879): 285–86.

72. Hospitalier, "Les horloges pneumatiques," 407.

73. Ibid., 409.

74. Lionel Hauser to Marcel Proust, January 25, 1910, Marcel Proust, *Correspondance*, vol. X, ed. Philip Kolb (Paris: Plon, 1983), 36–39, here 36.

75. Brassaï, *Proust in the Power of Photography*, 17-27.

76. On these and the following biographical details, see Ronald Hayman, *Proust: A Biography* (London: Heinemann, 1990); and Céleste Albaret, *Monsieur Proust: Souvenirs recueillis par Georges Belmont* (Paris: Laffont, 1973).

77. Georges Mareschal, "Le théâtrophone," *La Nature* 20 (1892): 55–58, here 58.

5. TIME TO PUBLISH

1. Alexander von Humboldt to Emil du Bois-Reymond, January 8, 1851, *Briefwechsel zwischen Alexander von Humboldt und Emil du Bois-Reymond*, 112.

2. See Wolfgang Pircher, "Gleichzeitigkeit," *Zeit und Geschichte: Kulturgeschichtliche Perspektiven*, ed. Erhard Chvojka, Andreas Schwarcz, and Klaus Thien (Vienna: Oldenbourg, 2002), 44–58.

3. Marcel Proust, *By Way of Sainte-Beuve*, trans. Sylvia Townsend Warner (London: Chatto & Windus, 1958), 47.

4. Ibid.

5. Ibid., 48 and 49 [modified].

6. Helmholtz to du Bois-Reymond, January 15, 1850, *Dokumente*, 90.

7. We may note in passing that only the publication in Müller's *Archiv* reproduces the complete title of the manuscript. The two other journals drop the line "Preliminary Report" and limit themselves to "On the Propagation Speed of Nerve Stimulations."

8. Du Bois-Reymond to Helmholtz, March 19, 1850, *Dokumente*, 92.

9. Ibid.

10. Helmholtz to du Bois-Reymond, April 22, 1850, ibid., 96–97.

11. Etienne-Jules Marey, *Animal Mechanism: A Treatise on Terrestrial and Aerial Locomotion* (London: King, 1874), 42; and *La méthode graphique dans les sciences expérimentales et particulièrement en physiologie et en médecine* (Paris: Masson, 1878), 145–46. See also Thomas Schestag, "Wiedergefunden: 'du temps perdu,'" 84–85; and Berz, *08/15*, 367.

12. Werner Siemens, *Lebenserinnerungen* (Berlin: Springer, 1892), 85.

13. Crosland, *Science under Control*, 248.

14. Humboldt to du Bois-Reymond, January 18, 1850, in *Briefwechsel zwischen Alexander von Humboldt und Emil du Bois-Reymond*, 101.

15. Du Bois-Reymond to Helmholtz, March 19, 1850, *Dokumente*, 92.

16. Helmholtz, "Vorläufiger Bericht," 71–72.

17. Helmholtz, "Note," 204.

18. Helmholtz, "Vorläufiger Bericht," 73.

19. Helmholtz, "Note," 205.

20. Helmholtz, "Vorläufiger Bericht," 72.

21. Helmholtz, "Note," 204.

22. Helmholtz, "Vorläufiger Bericht," 72–73.

23. Helmholtz, "Note," 205.

24. Alexander von Humboldt, *Kosmos: Entwurf einer physischen Weltbeschreibung*, vol. 3, (Stuttgart: Cotta, 1850), 93–96.

25. Humboldt to Helmholtz, February 12, 1850, qtd. Königsberger, *Hermann von Helmholtz*, vol. 1, 118.

26. Ibid.

27. Du Bois-Reymond to Helmholtz, March 19, 1850, *Dokumente*, 92.

28. Emil du Bois-Reymond, "Über die Geschichte der Wissenschaft," 1872, *Reden*, vol. 1, 431–40, esp. 432–36. On du Bois-Reymond's notion of the history of science, see Hans-Jörg Rheinberger, *On Historicizing Epistemology*, trans. David Fernbach (Stanford, CA: Stanford University Press, 2010), 5–6.

29. Du Bois-Reymond to Helmholtz, March 19, 1850, *Dokumente*, 92.

30. Carlo Matteucci, "Nouvelles expériences sur l'arc voltaïque," *Comptes rendus hebdomadaires des séances de l'Académie des sciences* 30 (1850): 201–4.

31. D. and T., "Feuilleton du National: 27 février. Académie des Sciences," *Le National*, February 27, 1850: 1.

32. Jean de La Fontaine, "Simonides Saved by the Gods," *The Complete Fables*, trans. Norman R. Shapiro (Urbana: University of Illinois Press, 2007), 17–19, here 17 [modified].

33. D. and T., "Feuilleton du National," 1.

34. Ibid.

35. See Helmholtz to du Bois-Reymond, April 22, 1850, *Dokumente*, 97: "Here, the *National* with its mocking article does not exist."

36. Du Bois-Reymond to Helmholtz, March 19, 1850, ibid., 93.

37. Helmholtz to A. F. J. Helmholtz, March 29, 1850, qtd. Königsberger, *Hermann von Helmholtz*, vol. 1, 121.

38. Anonymous, "Feuilleton du National: 30 mai. Académie des Sciences," *Le National*, May 30, 1849: 1.

39. Léon Foucault, "Feuilleton du Journal des débats du 1er juin 1849: Académie des Sciences: Séances des 21 et 28 mai," *Journal des débats politiques et littéraires*, June 1, 1849: 1. See also the German translation of this article in *Briefwechsel zwischen Alexander von Humboldt und Emil du Bois-Reymond*, appendix, document 4, 165–66.

40. Anonymous, "Séance de l'Académie des Sciences: Influence de la volonté sur l'électro-magnétismem," *La Lancette Française: Gazette des Hôpitaux civils et militaires*, July 5, 1849: 311. See also the German translation of this article in *Briefwechsel zwischen Alexander von Humboldt und Emil du Bois-Reymond*, appendix, document 6, 168.

41. Du Bois-Reymond to Helmholtz, March 19, 1850, *Dokumente*, 93.

42. On this topic, see Gabriel Finkelstein's important article "M. du Bois-Reymond Goes to Paris." In addition, see Eugenie Rosenberger, *Felix du Bois Reymond, 1782–1865* (Berlin: Meyer & Jessen, 1912), 282–86; and Wilhelm Haberling, "du Bois-Reymond in Paris 1850," *Deutsche medizinische Wochenschrift* 52 (February 5, 1920): 250–52.

43. Du Bois-Reymond to Helmholtz, March 19, 1850, *Dokumente*, 93; and Helmholtz to du Bois-Reymond, April 5, 1850, ibid., 94.

44. Emil du Bois-Reymond, Travel diary, Emil du Bois-Reymond Papers, Staatsbibliothek zu Berlin, Preußischer Kulturbesitz, Box 1, Folder 7, No. 1, 2–4. On "Mademoiselle Prudence," see Auguste Lassaigne, *Mémoires d'un magnétiseur, contenant la biographie de la somnambule Prudence Bernard* (Paris: Baillière, 1851). On September 15, 1850, Prudence was examined in a session of the Academy of Science in Milan. See Lassaigne, 79–89, and, more generally, Alan Gauld, *A History of Hypnotism* (Cambridge: Cambridge University Press, 1992), 169–70.

45. Du Bois-Reymond to Helmholtz, March 18, 1851, *Dokumente*, 106. Siemens remained in Paris until April 26. See Werner Siemens, "Sur la télégraphie électrique," *Comptes rendus hebdomadaires des séances de l'Académie des Sciences* 30 (1850): 434–37; "Sur la télégraphie électrique," *Annales de Chimie et de Physique* 29 (1850): 385–430; and *Mémoire sur la télégraphie électrique* (Berlin, 1851). For Siemens's view, see his letter of April 27 to his brother Wilhelm Siemens, in which he laments the fact that the description of his telegraph is published "without any drawings" (Conrad Matschoß, ed., *Werner*

Siemens: Ein kurzgefaßtes Lebensbild nebst einer Auswahl seiner Briefe, vol. 1
[Berlin: Springer, 1916], appendix, 79).

46. Du Bois-Reymond to Helmholtz, August 25, 1850, *Dokumente*, 100.

47. Du Bois-Reymond, Lecture in French, Emil du Bois-Reymond
Papers, Staatsbibliothek zu Berlin, Preußischer Kulturbesitz, Box 1, Folder 7,
Nos. 3, 4.

48. Du Bois-Reymond to Helmholtz, August 25, 1850, *Dokumente*, 100.

49. Du Bois Reymonnd to Carl Ludwig, April 9, 1850, *Zwei große
Naturforscher des 19. Jahrhunderts: Ein Briefwechsel zwischen Emil du Bois-
Reymond und Karl Ludwig*, ed. Estelle du Bois-Reymond (Leipzig: Barth,
1927), 88–89.

50. Du Bois-Reymond to Helmholtz, August 25, 1850, *Dokumente*, 100.

51. Du Bois-Reymond, Travel diary, 3–5. On Fizeau, see Jan Frercks,
"Creativity and Technology in Experimentation: Fizeau's Terrestrial Determination of the Speed of Light," *Centaurus* 42 (2000): 249–87. On Wheatstone (and Helmholtz) see Siegert, *Passage des Digitalen*, 357–68.

52. See the excerpts compiled by Marcel Verdet in Hermann Helmholtz,
"Mémoire sur la contraction des muscles de la vie animale et sur la vitesse de
propagation de l'action nerveuse," *Annales de chimie et de physique* 43, no. 3 (1855):
367–79; and Jules Béclard, *Traité élémentaire de physiologie humaine* (Paris:
Labé, 1855), 760–62.

6. MESSAGES FROM THE BIG TOE

1. Hermann Helmholtz, "The Facts in Perception," 1878, *Epistemological
Writings: The Paul Hertz/Moritz Schlick Centenary Edition of 1921*, trans.
Malcolm F. Lowe, ed. Robert S. Cohen and Yehuda Elkana (Dordrecht:
Reidel, 1977), 115–63, here 123.

2. Helmholtz to du Bois-Reymond, April 5, 1850, *Dokumente*, 94. Already
by the end of March, Helmholtz had told his father about this extension
of his research; see Helmholtz to A. F. J. Helmholtz, March 29, 1850, qtd.
Königsberger, *Hermann von Helmholtz*, vol. 1, 120.

3. Helmholtz to du Bois-Reymond, April 11, 1850, *Dokumente*, 112.

4. Helmholtz to du Bois-Reymond, April 5, 1850, ibid., 94

5. Ibid.

6. Helmholtz, "On the Methods," 324.

7. Helmholtz to du Bois-Reymond, April 5, 1850, *Dokumente*, 94.

8. Ibid.

9. Ibid., 94–95.

10. Helmholtz to A. F. J. Helmholtz, March 29, 1850, qtd. Königsberger,
Hermann von Helmholtz, vol. 1, 120.

11. Helmholtz, "On the Methods," 321.

12. Du Bois-Reymond to Helmholtz, mid-April 1852, *Dokumente*, 129.

13. Du Bois-Reymond to Helmholtz, February 9, 1852, ibid., 123.

14. Helmholtz to du Bois-Reymond, March 24, 1852, ibid., 128.

15. Helmholtz, "On the Methods,"323.

16. Ibid., 324.

17. Ibid. [modified].

18. Ibid., 320 [modified].

19. Christoph Hoffmann, "Helmholtz' Apparatuses: Telegraphy as Working Model of Nerve Physiology," *Philosophia Scientiae* 7, no. 1 (2003): 129–49. More generally, see Laura Otis, *Networking: Communicating with Bodies and Machines in the Nineteenth Century* (Ann Arbor: University of Michigan Press, 2001).

20. Helmholtz, "On the Methods," 314 [modified].

21. Ibid., 320

22. Ibid.

23. Ibid. See also Livy (Titus Livius), *Books I and II*, ed. and trans. Benjamin O. Foster (Cambridge: Harvard University Press, 1988), 325: "Drawing a parallel from this to show how like was the internal dissension of the bodily members to the anger of the plebs against the Fathers, he [Agrippa Menenius] prevailed upon the minds of his hearers."

24. Helmholtz, "On the Methods," 320 [modified].

25. Helmholtz, "Autobiographical Sketch," 388.

26. Helmholtz, "On the Methods," 324.

27. Ibid. [modified].

28. Ibid., my emphasis.

29. Ibid.

30. Ibid.

31. Ibid.

32. Ibid., 325 [modified].

33. Ibid., 321.

34. Ibid., 325.

35. Ibid., 314. On the history of the "circle of fire" example, see D. Anthony Larivière and Thomas M. Lennon, "The History and Significance of Hume's Burning Coal Example: Time, Identity, and Individuation," *Journal of Philosophical Research* 27 (2002): 511–26.

36. Helmholtz, "On the Methods," 321.

37. Ibid., 325.

38. Ibid.

39. Ibid.

40. Ibid. [amended and modified].

41. Ibid. On the problematic status of the whale in Melville's novel and in nineteenth-century comparative anatomy, see Robert Stockhammer, "Warum der Wal ein Fisch ist: Melvilles *Moby Dick* und die zeitgenössische Biologie," Bernhard J. Dotzler and Sigrid Weigel, eds., *"Fülle der combination": Literaturforschung und Wissenschaftsgeschichte* (Munich: Fink, 2005), 143–71.

42. Humboldt to du Bois-Reymond, July 29, 1851, in *Briefwechsel zwischen Alexander von Humboldt und Emil du Bois-Reymond*, 117.

43. Du Bois-Reymond to Helmholtz, April 11, 1851, *Dokumente*, 112.

44. Joseph Vogl, "Was ist ein Ereignis?" Peter Gente and Peter Weibel, eds., *Deleuze und die Künste* (Frankfurt: Suhrkamp, 2007), 67–83, here 75.

45. Hermann Helmholtz, "Mittheilung für die physikalische Gesellschaft in Berlin betreffend Versuche über die Fortpflanzungsgeschwindigkeit der Reizung in den sensiblen Nerven des Menschen," Archiv der Berlin-Brandenburgischen Akademie der Wissenschaften, Helmholtz Papers, NL Helmholtz 540 (1850), 1.

46. Ibid., 2. On this paper, see Klauß, "Die erste Mitteilung von H. Helmholtz."

47. Ibid., 3.

48. Ibid., 4.

49. Anonymous, "Im sechsten und siebten Jahre des Bestehens der physikalischen Gesellschaft wurden folgende Originaluntersuchungen von Mitgliedern in den Sitzungen vorgetragen," *Die Fortschritte der Physik in den Jahren 1850 und 1851* (Berlin: Reimer, 1855), 7–8, here 7.

7. THE RETURN OF THE LINE

1. Henri Michaux, *Emergences-Résurgences* (Geneva: Skira, 1993), 12.

2. Helmholtz to du Bois-Reymond, April 11, 1851, *Dokumente*, 111.

3. Ibid.

4. Hermann Helmholtz, "Über den Verlauf und die Dauer der durch Stromesschwankungen inducirten electrischen Ströme," *Bericht über die zur Bekanntmachung geeigneten Verhandlungen der Königl. Preuss. Akademie der Wissenschaften zu Berlin* (1851): 287–90; and "Ueber die Dauer und den Verlauf der durch Stromesschwankungen inducirten elektrischen Ströme," *Annalen der Physik* 83 (1851): 505–40.

5. Du Bois-Reymond to Helmholtz, May 16, 1851, *Dokumente*, 113.

6. Ibid.

7. Helmholtz to du Bois-Reymond, June 12, 1851, ibid., 114–15.

8. Helmholtz, "On the Methods," 321.

9. Ibid., 323 [modified].

10. Ibid. [modified].

11. Ibid.

12. Helmholtz to du Bois-Reymond, September 17, 1850, *Dokumente*, 106.

13. On this point, see Königsberger, *Hermann von Helmholtz*, vol. 1, 133–44.

14. Helmholtz, "On the Methods," 323.

15. Ibid.

16. Helmholtz to du Bois-Reymond, April 11, 1851, *Dokumente*, 111.

17. This is also the argument proffered by de Chadarevian, in "Graphical Method and Discipline" (283), who picks up on Olesko and Holmes (and Nicholas Jardine).

18. Helmholtz, "Messungen (Zweite Reihe)," 202.

19. Ibid., 206.

20. Ibid., 204.

21. Helmholtz, "Deuxième Note," 264.

22. Helmholtz, "Messungen (Zweite Reihe)," 210.

23. Ibid.

24. Ibid.

25. Ibid., 211.

26. Ibid.

27. Ibid., 212.

28. Soraya de Chadarevian, "Die Handschrift der Natur: Selbstschreibende Geräte in der Physiologie des 19. Jahrhunderts," *Der Druck des Wissens: Geschichte und Medium der wissenschaftlichen Publikation*, ed. Michael Cahn (Wiesbaden: Reichert, 1991), 21–22, here 22.

29. Helmholtz, "Messungen (Zweite Reihe)," 211, emphasis added.

30. Helmholtz, "Deuxième Note," 265.

31. Helmholtz, "Messungen (Zweite Reihe)," 213.

32. Ibid., 212.

33. Ibid., 214.

34. Ibid., 214–15.

35. Ibid., 215.

36. Ibid., 216.

37. Ibid.

38. Hermann Helmholtz, "Über die Geschwindigkeit einiger Vorgänge in Muskeln und Nerven," *Bericht über die zur Bekanntmachung geeigneten Verhandlungen der Königlich Preußischen Akademie der Wissenschaften zu Berlin* 19 (1854): 328–32.

39. Ibid., 332.

40. Ibid., 328 and 329. See also Müller, *Elements of Physiology*, 613–19 and 787–90; and Hermann Lotze, "Seele und Seelenleben," *Handwörterbuch der*

Physiologie mit Rücksicht auf physiologische Pathologie, 142–264 (e.g., 203). On this issue, see also Christoph Hoffmann, "Aufschub: Talbot, Helmholtz und das Ereignis der Latenzzeit," *Latenz: 40 Annäherungen an einen Begriff,* ed. Stefanie Diekmann and Thomas Khurana (Berlin: Kadmos, 2007), 31–34.

41. Schelske, "Neue Messungen der Fortpflanzungsgeschwindigkeit."

42. Ibid., 153 and 173.

43. Undated letter by Koenig to Donders, qtd. Franciscus Cornelis Donders, *Snelheid der Werking in 't Zenuwstelsel,* ca. 1865, Universiteitsmuseum Utrecht, Donders Papers, Do. 07.77, 4. See also Schmidgen, "The Donders Machine," 240.

44. On this point, see Schmidgen, "The Donders Machine," 243–52.

45. Emil du Bois-Reymond, "On the Time Required for the Transmission of Volition and Sensation through the Nerves" [April 13, 1866], *Notices of the Proceedings at the Meetings of the Members of the Royal Institution of Great Britain* 4 (1862–1866): 575–93.

46. Etienne-Jules Marey, *Du mouvement dans les fonctions de la vie* (Paris: Baillière, 1868), 418.

47. Ibid., 428.

48. Ibid. On cylinder recordings with tuning forks, see Marey, *Animal Mechanism,* 43, and *La méthode graphique,* 147. See also de Chadarevian, "Graphical Method and Discipline," 288, n. 60.

49. Du Bois-Reymond, "On the Time Required," 579. As for textbooks and manuals, see for example Marey, *Du mouvement dans les fonctions de la vie,* 415; and Conrad Eckhard, *Experimentalphysiologie des Nervensystems* (Gießen: Roth, 1867), 136.

50. On these contrivances, see Schmidgen, "Pictures, Preparations, and Living Processes."

51. Du Bois-Reymond, "Der physiologische Unterricht sonst und jetzt," 650–51.

52. Marey, *La méthode graphique,* 145–46; and "La méthode graphique dans les sciences expérimentales (1er article)," *Physiologie Expérimentale: Travaux du laboratoire de M. Marey* (1875): 123–64, here 151.

53. Du Bois-Reymond, "On the Time Required," 577.

54. Ibid., 588. Like Hirsch and Schelske, du Bois-Reymond here assumes that 26 to 30 meters per second was the correct figure for the propagation speed of nervous stimulations in humans. This contradicts the result Helmholtz had obtained, that is, 60 meters per second. In the 1860s and 1870s, the attempt to clarify this striking divergence led to numerous studies concerning the psychophysiology of time. They were carried out by Donders, de Jaager, Helmholtz, and others.

55. See, for example, Albert Londe, *La photographie moderne: Pratique et applications* (Paris: Masson, 1888), 272; and Frédéric Dillaye, *La théorie, la pratique et l'art en photographie* (Paris: Librairie Illustrée, 1891), 99.

56. Gilles Deleuze and Félix Guattari, *What is Philosophy?* trans. Hugh Tomlinson and Graham Burchell (New York: Columbia University Press, 1994), 118. Du Bois-Reymond mentions instantaneous photography only later. See his "Naturwissenschaft und bildende Kunst: Zur Feier der Leibniz-Sitzung der Akademie der Wissenschaften zu Berlin am 3. Juli 1890 gehaltene Rede," *Reden*, vol. 2, 390–425, here 409. The book that du Bois-Reymond quotes in this context, Joseph M. Eder's *Die Moment-Photographie in ihrer Anwendung auf Kunst und Wissenschaft* (2nd rev. ed. [Halle: Knapp, 1886]) also contains a comparative table of velocities (11). On du Bois-Reymond's lecture and the history of cinematography, see Friedrich Kittler, "Der Mensch, ein betrunkener Dorfmusikant," Helmar Schramm et al., eds., *Bühnen des Wissens: Interferenzen zwischen Wissenschaft und Kunst* (Berlin: Dahlem University Press, 2003), 300–18.

57. Du Bois-Reymond, "On the Time Required," 589.

58. Ibid.

59. Paul Virilio, *Polar Inertia*, trans. Patrick Camiller (London: Sage, 2000).

CONCLUSION

1. Proust, *Finding Time Again*, 197.

2. Gilles Deleuze, *Proust and Signs*, trans. Richard Howard (Minneapolis: University of Minnesota Press, 2000), 4.

3. Ibid.

4. François Jacob, *The Logic of Life: A History of Heredity*, trans. Betty E. Spillmann (Princeton, NJ: Princeton University Press, 1993), 153.

5. Hermann Helmholtz, "The Facts in Perception," 122. However, see his text of 24 years ealier, "Ueber die Natur der menschlichen Sinnesempfindungen," *Königsberger naturwissenschaftliche Unterhaltungen* 3 (1854): 1–20, here 19: "Sensations of light and color are just symbols for relations of reality."

6. The entire first part of Marey's book *La méthode graphique* (1–106) highlights this distinction between representation (*représentation*) and recording (*inscription*).

7. Proust, *Finding Time Again*, 188.

8. Hermann von Helmholtz, *Helmholtz's Treatise on Physiological Optics*, 1867, vol. 3, *The Perceptions of Vision*, ed. and trans. James P. C. Southall, 1925 (New York: Dover, 1962), 30–31.

This bibliography provides an overview of Helmholtz's publications (including translations) and of the secondary literature on his time experiments and on the graphical method. It thus does not list all the titles already cited in the text and documented in the notes. Each section of the bibliography is arranged chronologically. Ordered this way, the titles delineate the development of research on Helmholtz and of the engagement with the manifold traces of time in experimental physiology on the part of historians of science and of art.

PUBLICATIONS BY HELMHOLTZ (TRANSLATED TITLES AND SHORT TITLES USED IN THE TEXT ARE IN BRACKETS)

"Vorläufiger Bericht über die Fortpflanzungs-Geschwindigkeit der Nervenreizung [Preliminary Report on the Propagation Speed of Nerve Stimulations]" ["Preliminary Report"], *Archiv für Anatomie, Physiologie und wissenschaftliche Medicin* (1850): 71–73.

"Note sur la vitesse de propagation de l'agent nerveux dans les nerfs rachidiens [Note on the Propagation Speed of the Nervous Agent in the Spinal Nerves]" ["Note"], *Comptes rendus hebdomadaires des séances de l'Académie des sciences* 30 (1850): 204–8.

"Messungen über den zeitlichen Verlauf der Zuckung animalischer Muskeln und die Fortpflanzungsgeschwindigkeit der Reizung in den Nerven [Measurements Concerning the Temporal Course of the Contraction in Animal Muscles and the Propagation Speed of Stimulations in the Nerves]" ["Measurements"], *Archiv für Anatomie, Physiologie und wissenschaftliche Medicin* (1850): 276–364.

"Ueber die Methoden, kleinste Zeittheile zu messen, und ihre Anwendung für physiologische Zwecke," *Koenigsberger naturwissenschaftliche Unterhaltungen* 2 (1850): 169–89; "On the Methods of Measuring Very Small Portions of Time, and Their Application to Physiological Purposes" ["On the Methods"],

The London, Edinburgh and Dublin Philosophical Magazine and Journal of Science 4 (1853): 313–25.

"Ueber die Dauer und den Verlauf der durch Stromesschwankungen inducirten elektrischen Ströme [On the Course and Duration of Electrical Currents Induced by Fluctuations of Currents]," *Annalen der Physik und Chemie* 83 (1851): 505–40.

"Deuxième Note sur la vitesse de propagation de l'agent nerveux [Second Note on the Propagation Speed of the Nervous Agent]" ["Second Note"], *Comptes rendus hebdomadaires des séances de l'Académie des sciences* 33 (1851): 262–65.

"Messungen über Fortpflanzungsgeschwindigkeit der Reizung in den Nerven (Zweite Reihe) [Measurements Concerning the Propagation Speed of Irritations in the Nerves (Second Series)]" ["Measurements (Second Series)"], *Archiv für Anatomie, Physiologie und wissenschaftliche Medicin* (1852): 199–216.

"Ueber die Geschwindigkeit einiger Vorgänge in Muskeln und Nerven [On the Speed of Some Processes in Muscles and Nerves]," *Bericht über die zur Bekanntmachung geeigneten Verhandlungen der Königlich Preussischen Akademie der Wissenschaften zu Berlin* (1854): 328–32.

ON HELMHOLTZ'S TIME EXPERIMENTS

M'Kendrick, John Gray, "Helmholtz in Königsberg—Measurement of the Rapidity of the Nervous Impulse," *Hermann Ludwig Ferdinand von Helmholtz* (New York: Longmans, Green & Co., 1899), 58–70.

Königsberger, Leo, "Helmholtz als Professor der Physiologie in Königsberg vom Sommer 1849 bis Michaelis 1855," *Hermann von Helmholtz*, vol. 1 (Braunschweig: Vieweg, 1902), 111–256.

Boring, Edwin G., "Psychophysiology in the First Half of the Nineteenth Century: The Volicity and Conduction of the Nervous Impulse," *A History of Experimental Psychology* (New York: Century, 1935), 42–46.

Hoff, Hebbel E., and Leslie A. Geddes, "Graphic Registration before Ludwig: The Antecedents of the Kymograph," *Isis* 50, no. 159 (1959): 5–21.

Hoff, Hebbel E., and Leslie A. Geddes, "Ballistics and the Instrumentation of Physiology: The Velocity of the Projectile and of the Nerve Impulse," *Journal of the History of Medicine and Allied Sciences* 15, no. 2 (1960): 133–46.

Hoff, Hebbel E., and Leslie A. Geddes, "The Technological Background of Physiological Discovery: Ballistics and the Graphic Method," *Journal of the History of Medicine and Allied Sciences* 15, no. 4 (1960): 345–63.

Blasius, Wilhelm, "Die Bestimmung der Leitungsgeschwindigkeit im Nerven durch Hermann v. Helmholtz am Beginn der naturwissenschaftlichen Ära der Neurophysiologie," *Von Boerhaave bis Berger: Die Entwicklung der kontinentalen Physiologie im 18. und 19. Jahrhundert mit besonderer Berücksichtigung der Neurophysiologie*, ed. Karl E. Rothschuh (Stuttgart: Fischer, 1964), 71–84.

Olesko, Kathryn M., and Frederic L. Holmes, "Experiment, Quantification, and Discovery: Helmholtz's Early Physiological Researches, 1843–1850," *Hermann von Helmholtz and the Foundations of Nineteenth-Century Science*, ed. David Cahan (Berkeley: University of California Press, 1993), 50–108.

Chadarevian, Soraya de, "Graphical Method and Discipline: Self-Recording Instruments in Nineteenth-Century Physiology," *Studies in History and Philosophy of Science* 24 (1993): 267–91.

Holmes, Frederic L., and Kathryn M. Olesko, "The Images of Precision: Helmholtz and the Graphical Method in Physiology," *The Values of Precision*, ed. Norton Wise (Princeton, NJ: Princeton University Press, 1994), 198–221.

Brain, Robert M., and M. Norton Wise, "Muscles and Engines: Indicator Diagrams and Helmholtz's Graphical Methods," *Universalgenie Helmholtz: Rückblick nach 100 Jahren*, ed. Lorenz Krüger (Berlin: Akademie, 1994), 124–45.

Klauß, Klaus, "Die erste Mitteilung von H. Helmholtz an die Physikalische Gesellschaft über die Fortpflanzungsgeschwindigkeit der Reizung in den sensiblen Nerven des Menschen," *NTM* (N.S.) 2 (1994): 89–96.

Debru, Claude, "Helmholtz and the Psychophysiology of Time," *Science in Context* 14, no. 3 (2001): 471–92.

Nicolas, Serge, and Ludovic Ferrand, "La psychométrie sensorielle au XIXe siècle," *Psychologie et Histoire* 2 (2001): 148–73.

Schmidgen, Henning, "Of Frogs and Men: The Origins of Psychophysiological Time Experiments, 1850–1865," *Endeavour* 26, no. 4 (2002): 142–48.

Finger, Stanley, and Nicholas J. Wade, "The Neuroscience of Helmholtz and the Theories of Johannes Müller: Part 1: Nerve Cell Structure, Vitalism, and the Nerve Impulse," *Journal of the History of the Neurosciences* 11 (2002): 136–55.

Schmidgen, Henning, "Die Geschwindigkeit von Gedanken und Gefühlen: Die Entwicklung psychophysiologischer Zeitmessungen, 1850–1865," *NTM* (N.S.) 12 (2004): 100–115.

Wise, M. Norton, "What Can Local Circulation Explain? The Case of Helmholtz's Frog-Drawing-Machine in Berlin," *Journal for the History of Science and Technology* 1 (2007): 15–73.

Wise, M. Norton, *Neo-Classical Aesthetics of Art and Science: Hermann Helmholtz and the Frog-Drawing Machine* (Uppsala: Uppsala University, 2008).

ON THE "GRAPHICAL METHOD"

Panconcelli-Calzia, Giulio, "Zur Geschichte des Kymographions," *Zeitschrift für Laryngologie, Rhinologie, Otologie und ihre Grenzgebiete* 26 (1935): 196–207.

Funkenhouser, H. Gray, "Historical Development of the Graphical Representation of Statistical Data," *Osiris* 3 (1937): 269–404.

Sigfried Giedion, *Mechanization Takes Command: A Contribution to Anonymous History* (Oxford: Oxford University Press, 1948).

Hoff, Hebbel E., Leslie A. Geddes, and Roger Guillemin, "The Anemograph of Ons-en-Bray: An Early Self-Registering Predecessor of the Kymograph with Translations of the Original Description and a Biography of the Inventor," *Journal of the History of Medicine and Allied Sciences* 12, no. 4 (1957): 424–48.

Hoff, Hebbel E., and Leslie A. Geddes, "The Beginning of Graphic Recording," *Isis* 53 (1962): 287–310.

Frizot, Michel, ed., *E. J. Marey 1830–1904: La photographie du mouvement* (Paris: Centre national d'art et de culture Georges Pompidou, 1977).

Borell, Merilley, "Extending the Senses: The Graphic Method," *Medical Heritage* 2 (1986): 114–21.

Dagognet, François, *Etienne-Jules Marey: La passion de la trace* (Paris: Hazan, 1987).

Frank, Robert G., "The Telltale Heart: Physiological Instruments, Graphic Methods, and Clinical Hopes, 1854–1914," *The Investigative Enterprise: Experimental Physiology in Nineteenth-Century Medicine*, ed. William Coleman and Frederic L. Holmes (Berkeley: University of California Press, 1988), 211–90.

Rabinbach, Anson, *The Human Motor: Energy, Fatigue, and the Origins of Modernity* (Berkeley: University of California Press, 1990).

Chadarevian, Soraya de, "Die Handschrift der Natur: Selbstschreibende Geräte in der Physiologie des 19. Jahrhunderts," *Der Druck des Wissens: Geschichte und Medium der wissenschaftlichen Publikation*, ed. Michael Cahn (Wiesbaden: Reichert, 1991).

Braun, Marta, *Picturing Time: The Work of Etienne-Jules Marey (1830–1904)* (Chicago: University of Chicago Press, 1992).

Daston, Lorraine, and Peter Galison, "The Image of Objectivity," *Representations* 40 (1992): 81–128.

Cartwright, Lisa, *Screening the Body: Tracing Medicine's Visual Culture* (Minneapolis: University of Minnesota Press, 1995).

Hankins, Thomas L., and Robert J. Silverman, "Science Since Babel: Graphs, Automatic Recording Devices, and the Universal Language of Instruments," *Instruments and the Imagination* (Princeton: Princeton University Press, 1995), 113–47.

Douard, John W., "E.-J. Marey's Visual Rhetoric and the Graphic Decomposition of the Body," *Studies in History and Philosophy of Science, Part A* 26, no. 2 (1995): 175–204.

Brain, Robert M., *The Graphic Method: Inscription, Visualization, and Measurement in Nineteenth-Century Science and Culture*, PhD dissertation, University of California, Los Angeles, 1996.

Brain, Robert M., "Standards and Semiotics," *Inscribing Science: Scientific Texts and the Materialities of Communication*, ed. Timothy Lenoir (Stanford: Stanford University Press, 1998), 249–84.

Hankins, Thomas L., "A Particular History of Graphs," *Isis* 90 (1999): 50–80.

Mannoni, Laurent, *Etienne-Jules Marey: La mémoire de l'œil* (Milan: Mazzotta, 1999).

Brain, Robert M., "Representation on the Line: Graphic Recording Instruments and Scientific Modernism," *From Energy to Information: Representation in Science and Technology, Art and Literature*, ed. Bruce Clarke and Linda Dalrymple Henderson (Stanford: Stanford University Press, 2002), 155–77.

Schäffner, Wolfgang, "Mechanische Schreiber. Jules Etienne Mareys Aufzeichnungsmaschinen," *Europa: Kultur der Sekretäre*, ed. Bernhard Siegert and Joseph Vogl (Zurich: Diaphanes, 2003), 221–35.

Prodger, Phillip, ed., *Time Stands Still: Muybridge and the Instantaneous Photography Movement* (New York: Cantor Center for Visual Arts at Stanford University/Oxford University Press, 2003).

Didi-Huberman, Georges, and Laurent Mannoni, *Mouvements de l'air: Etienne-Jules Marey, photographe des fluides* (Paris: Gallimard, 2004).

Borck, Cornelius, *Hirnströme: Eine Kulturgeschichte der Elektroenzephalographie* (Göttingen: Wallstein, 2005).

Borck, Cornelius, and Armin Schäfer, eds., *Psychographien* (Zurich and Berlin: Diaphanes, 2005).

Schmidgen, Henning, "The Donders Machine: Matter, Signs, and Time in a Physiological Experiment, c. 1865," *Configurations* 13, no. 2 (2005/2007): 211–56.

Felsch, Philipp, *Laborlandschaften: Physiologische Alpenreisen im 19. Jahrhundert* (Göttingen: Wallstein, 2007).

Schäffner, Wolfgang, "Bewegungslinien: Analoge Aufzeichnungsmaschinen," Michael Franz, Wolfgang Schäffner, Bernhard Siegert, and Robert Stockhammer, eds., *Electric Laokoon: Zeichen und Medien, von der Lochkarte zur Grammatologie* (Berlin: Akademie, 2007), 130–45.

Métraux, Alexandre, "Wahrnehmungsmaschinen: Wie Charles Cros das Sehen digitalisierte," in *Electric Laokoon*, 153–71.

Amelunxen, Hubertus von, Dieter Appelt, and Peter Weibel, eds., *Notation: Kalkül und Form in den Künsten* (Berlin: Akademie der Künste; Karlsruhe: Zentrum für Kunst und Medientechnologie, 2008).

Rieger, Stefan, *Schall und Rauch: Eine Mediengeschichte der Kurve* (Frankfurt: Suhrkamp, 2009).

forms of living

Stefanos Geroulanos and Todd Meyers, *series editors*

Georges Canguilhem, *Knowledge of Life*. Translated by Stefanos Geroulanos and Daniela Ginsburg, Introduction by Paola Marrati and Todd Meyers.

Henri Atlan, *Selected Writings: On Self-Organization, Philosophy, Bioethics, and Judaism*. Edited and with an Introduction by Stefanos Geroulanos and Todd Meyers.

Catherine Malabou, *The New Wounded: From Neurosis to Brain Damage*. Translated by Steven Miller.

François Delaporte, *Chagas Disease: History of a Continent's Scourge*. Translated by Arthur Goldhammer, Foreword by Todd Meyers.

Jonathan Strauss, *Human Remains: Medicine, Death, and Desire in Nineteenth-Century Paris*.

Georges Canguilhem, *Writings on Medicine*. Translated and with an Introduction by Stefanos Geroulanos and Todd Meyers.

François Delaporte, *Figures of Medicine: Blood, Face Transplants, Parasites*. Translated by Nils F. Schott, Foreword by Christopher Lawrence.

Juan Manuel Garrido, *On Time, Being, and Hunger: Challenging the Traditional Way of Thinking Life*.

Pamela Reynolds, *War in Worcester: Youth and the Apartheid State*.

Vanessa Lemm and Miguel Vatter, eds., *The Government of Life: Foucault, Biopolitics, and Neoliberalism*.

Henning Schmidgen, *Bruno Latour in Pieces: An Intellectual Biography*. Translated by Gloria Custance.

Veena Das, *Affliction: Health, Disease, Poverty*.

Kathleen Frederickson, *The Ploy of Instinct: Victorian Sciences of Nature and Sexuality in Liberal Governance*.

Roma Chatterji (ed.), *Wording the World: Veena Das and Scenes of Instruction*.

Jean-Luc Nancy and Aurélien Barrau, *What's These Worlds Coming To?* Translated by Travis Holloway and Flor Méchain. Foreword by David Pettigrew.